Contents

Preface by General Editor *v*

Author's introduction *vii*

1 Science education in the primary and middle school years

Introduction *1*
Content and process goals in primary science *5*
Early scientific development *7*
Assessment – a challenge *10*

2 Purposes and methods of assessing children

Meanings *12*
The 'why', 'what' and 'how' of assessment *13*
Various reasons 'why' *14*
General points about 'how' *15*
Examples of using the features *19*

3 Day to day assessment of scientific development – 1

Introduction *22*
The role of assessment in matching *22*
Information required as feedback *24*
Possible methods of assessment *27*

4 Day to day assessment of scientific development – 2 (Examples)

Assessing by observation in normal activities – an example *30*
Assessing by observation in special situations – examples *33*
The process of assessing by observation *35*
Selection of criteria *36*
Examples of criteria *38*

5 Summing up achievement

Introduction *40*
Purposes of summary assessment *41*
The information required from summary assessment *42*
Methods *42*
National surveys in science – the work of the Assessment of Performance Unit *54*

6 Keeping and using records of assessment

Are records really necessary? *59*
Types of record *60*
Records for different purposes *64*
Reviewing and improving records *65*
Interpretation of records *67*

7 Evaluating opportunities for learning

Introduction *69*
The meaning of evaluation *70*
Information about learning opportunities *71*
Criteria for evaluation of learning opportunities *74*
Applying the criteria *77*
The value of evaluation *78*

References *80*

Index *82*

Preface

This book is one of a series on assessment in school. The underlying theme of all the books is that there is a vital interaction between the art of teaching and the craft of assessing or evaluating. When I was a young teacher the first piece of advice given to me was 'Teach and test'. The principles involved in measurement in education are relatively simple but they sometimes appear complex for two rather different reasons. First, the context of the assessment is often complicated because schools, classrooms, pupils and teachers are immensely varied, and second, there is an apparent necessity to use statistical procedures as an essential ingredient in the measurement process. In our series of books we have concentrated on the principles of assessment and therefore the books are relatively non-technical. We have also made sure that the points about assessment, important as they are, are made subsidiary to the curriculum and teaching aspects of the subject. In short, the basic idea is that the act of measurement helps to concentrate the mind and to make precise those aspects of the teaching process which should be made precise. All teachers, creative or otherwise, do well to put their ideas and methods to the test and discipline of assessment procedures. On the other hand, it is all too easy to allow testing and examining procedures to stifle the creative heart of a subject. All the authors in this series are determined that this should not happen. So although the books are about assessment, a great deal is written about the content and method of teaching. Each book can, of course, be read independently, but the series is planned as a coherent whole and there will be many readers who will find value in reading more than one book in the series.

The relationship between teaching and assessment is particularly important in science, especially in the primary and middle school. The place of science in the curriculum is not yet secure in the earlier years of schooling and knowledge about science amongst non-specialist teachers is understandably patchy. Add to this the fact that there are relatively few good science tests available and it is easy to see the delicate nature of the situation. Good science teaching is hard to achieve and can easily be stifled by bad testing. That many teachers have an inadequate knowledge of science is well-known, but more serious, however, is the less obvious point that the underlying philosophy and principles of science are not quite the same as the non-expert might think. Wynne Harlen has therefore found it necessary to write a good deal about science teaching itself before discussing the evaluation and assessment aspects of the topic. Her experience and background make her an ideal author for this book. Beginning as the evaluator for the well-known Science 5–13 project (after evaluating the less well-known Oxford Primary Science Project) she developed interesting ideas about evaluation which led to her directing her own project on Progress and Learning in Science, and she now works at Chelsea College, University of London, on the Assessment of Performance Unit work in monitoring the national scene in science achievement.

Throughout her professional career Wynne Harlen has wrestled with this fundamental problem concerned with the relationship between teaching and assessing. As a physicist, she quite naturally began in the belief that psychometric methods were truly scientific and expected to use such methods in evaluating Science 5–13. She quickly found it necessary to use more eclectic methods and employed methods of evaluation which have come to be called 'illuminative'. She now believes, as I do, that evaluation should be subservient to the dictates of the teaching situation. In this book she first describes the essentials of science itself in the school context, and then makes proposals for the proper measure of progress in science using the skills and knowledge of the teacher, as well as suggesting which tests are available and worth using.

Readers of this book will find it helpful to read the companion volume in this series by Bentley and Malvern on *Mathematics*. Taken together, the two books will enable teachers to improve their knowledge of the two subjects as well as improving their knowledge about how to make proper and accurate assessments.

Jack Wrigley
General Editor

Author's introduction

Assessment in education has been criticized for interfering with the process of learning, the analogy being that of the gardener constantly pulling up his plants to see if the roots are growing. There is some truth in this, particularly if there is too much assessment of the wrong kind, but it also distorts reality to make a point. Gardeners do have to find out if their plants are growing and they do this, not by uprooting them, but by careful observation with a knowledgeable eye, so that they can give water and food at the right times and avoid either undernourishment or over-watering. Similarly, assessing children's educational development is an essential part of fostering that development, of teaching, so that both short and long-term planning of activities can match the stages reached and provide optimum opportunities for further learning.

To extend the gardening metaphor a little further, it is useful to emphasize that assessment requires a trained eye. The gardener who does not know what size and shape a plant is to be and how quickly it is expected to grow will not know what signs to look for, and may mistake a condition which is quite normal for one which requires remedial action, or vice versa. We have to know something about the development taking place in order to interpret what we find when we assess it. Children's learning is much more complex than plant growth, of course, and the conditions which promote and inhibit it are much less clear. This complexity serves to emphasize the importance of knowing how to recognize development when it happens. So a book about assessment should give a clear idea of the nature of the learning being assessed, in this case learning in science in the primary and middle school years. The first chapter attempts to do this through looking at the classroom and children's activities and then at the goals and objectives of early science education stated or reflected in various published teaching materials. Some observations about content and process goals are made en route.

In order to use assessment to help teaching it is necessary to appreciate the different purposes that assessment can serve and the strengths and weaknesses of the various ways in which it can be carried out. Chapter Two reviews briefly several types of purpose and suggests how the choice of what is assessed and of the methods used might vary from one purpose to another. This short chapter is the most theoretical of the book, but it introduces ideas used in later chapters. One of its main functions is to provide a way of taking assessment methods apart (pp 15–19) so that they can be analysed, criticized and more readily related to their purpose. Thus it is worthwhile for the reader to dip into it before going on to the more practically oriented chapters, though a more detailed look at specific points could be delayed until later.

Assessment need not necessarily focus only upon the children. To improve learning it is as relevant to assess the opportunities for learning provided by the teaching as it is to assess what has been learned. The remaining chapters of the book deal with both of these aspects, discussing and giving examples of methods which can be used to gather the relevant information.

Chapters Three, Four and Five concern the assessment of children for two important purposes; for improving day to day decisions about children's activities and for summing up children's achievement over a certain period of time. Chapter Five also includes a brief reference to assessment carried out by the APU to serve purposes quite different from those of teacher's own assessment.

The types of record which can be made of assessment and the interpretation and use of records are the topics of Chapter Six. Forms of record – from free comment to checklists to grades and test scores – have to be chosen to suit the nature of the information gathered and the purpose for which it is intended. Some criteria are suggested for evaluating the types of record which might be kept within a school. This leads to the subject of using records and taking action on the basis of the information they contain.

To make use of records it is not generally enough to know how the pupils performed but necessary to know something of what opportunities were provided for their learning. This point is taken further in Chapter Seven, which opens with a discussion of the meaning of evaluating opportunities for learning and attempts to relate this process to the assessment of children's learning. Some methods are outlined for evaluating the opportunities provided through the content, methods and organization of classroom activities. The importance of defining criteria for evaluation is emphasized, and some examples are given of ways of expressing and of using criteria for this purpose.

This book has been written with practising teachers in mind, particularly head teachers and those with special interest or responsibility for science. No book can be tailor-made to the needs of each of its readers. For some there will be a great deal more in the book than they require; for others too little detail of some kinds and too much of other kinds. It is not necessary to suppose that the book should be read from cover to cover or not at all. One way of sifting through the ideas, and extending them, is to use sections as discussion material among groups of teachers. It is hoped that the book will be used at times in this way, both in schools and in teachers' courses, since then teachers' own views are added to those presented and are made available to others.

Wynne Harlen, University of London
April 1982

1 Science education in the primary and middle school years

Introduction

Before we begin to talk about assessing science in the primary and middle school years it is necessary to be clear about what it is that we are assessing. What is the meaning of science in these early years and what do children learn from it? To answer the first of these questions we can do no better than to step into a school where science is, and has been for many years, a significant part of the curriculum.

In a first year junior class (seven and eight year olds), one topic concerns the weather. This is not the only science topic going on at the time in the class, since it has been concerned with the keeping of weather records over a long period, but it needs regular, although not extensive, attention. The usual types of records have been kept, involving using thermometers for air and ground temperature, a rain gauge, a maximum and minimum thermometer and a wind vane. These instruments have been investigated and the children have some idea how they work and know what the measurements mean. They have made thermometers from bottles and glass tubing, and they have poured water into the rain gauge and calculated what a certain depth means in terms of depth of water over the collecting area.

At some periods the weather project is a matter of routine record keeping but at certain intervals it becomes the centre of attention. There are times when there is some summary of the records; bar charts are drawn of the weekly rainfall over a period of half a term, a wind-rose constructed to record the wind directions, and line graphs drawn of the various temperatures which have been recorded. Groups work together making these charts and graphs and then display the result for the others to see. In group and class discussions the teacher helps them to make some attempt at looking for patterns in their records – do the lowest minimum temperatures go with the lowest day temperatures? Is there any association between rainfall and temperature? Was the wind blowing more often from one direction than from others? Invariably these discussions lead to ideas for further work. Can we see how good the BBC weather forecast is? What do we mean by 'how good'? On another occasion a simple and home-made device for measuring wind speed was added to the collection of instruments because some children had made the suggestion that the wind was always stronger from the west. Their records helped them to test this hypothesis. They found it was true at one time of the year, but not at another and this made them look for other differences in weather patterns as the year went by.

If we leave this class and eavesdrop on a second year junior class we find groups of children enjoying the challenge of seeing how many drawing pins they can put in boats made of plasticine before they sink. This activity was part of a series which started from a visit to see the launching of a ship in the city's dockyard. The visit had first been followed up by some work on pollution, since the dirty state of the shoreline had been one of the things to impress the children most. Then they had investigated the floating or sinking of various objects and materials. However, now their attention had turned to how heavy metal ships could float and they were finding out the shapes which could take most load. The teacher noted that the children were taking lumps of plasticine and comparing the boats they could make from them without any regard to starting with equal amounts of plasticine. Did the amount of plasticine make any difference? It was soon found that it did. Some childen were content, however, to take pieces which *looked* about the same, but the suggestion of one child to weigh the pieces led to other ideas of ways of making sure they *were* the same. Another child explained that if they used new plasticine there was no need to weigh since equal lengths would be the same because the width and depth were the same. They continued with renewed energy, after equalizing their lumps of plasticine, and noticeably paid attention to making sure other things were 'fair', for instance that they all used the same kind of drawing pins to load their boats.

Topics of work in this school often begin from visits. The teachers find these engage children's interest and start them off on activities which have real meaning to them i.e. they are genuinely curious about what they have seen and really want to find out about it. So if we move on to a third class (nine and ten year olds) we find one group busy investigating parachutes, another two groups testing paper planes and a further two persevering with the problems of jet propulsion in the form of a sausage-shaped balloon attached to a straw threaded on to a string stretched across the room. (The ideas came from the Science 5/13 Unit *Holes, Gaps and Cavities*, 1973)

These activities had all started from a visit to the airport. The first enthusiasm had been for making model planes, first out of paper and then out of balsa wood. This had involved discussion of the shape of an aerofoil and some activities through which they came to understand the reason for the shape. Some children brought planes they had made at home from kits of parts. They were keen to compare them and compete to find the best flyer. How would they judge this? What would be a fair test? Is it just horizontal distance that matters or the length of flight in the air? Or the length of time before the plane comes down? These matters had to be discussed and procedures for measurement agreed before the boys would risk their precious models in the competition. Similar consideration of variables was involved in the comparison of parachutes with and without holes. The area of the parachute and its load were also factors which these older children realized must be the same for a fair test. In the construction of the rocket it was the girls who came into their own. Finding a way both to hold a balloon inflated but ready for quick release and at the same time attach it to a straw threaded on the string involved tricky physical manipulation, careful planning, and team work. They got as far as seeing that if the balloon was fully inflated it travelled further than when it was not fully inflated, but exact measurements were not taken. They were satisfied to have made it work and obtained some evidence to support their idea that the more the balloon was blown up the further it would go.

In the fourth class that we shall visit the work had also begun from a visit, this time not very far away but to the 'wild' patch in the school garden. The teachers had been trying to improve the potential for science work in the school grounds, with some success. Part of the asphalt had been removed to create a grassed area on which a large tree trunk was laid. It not only made a useful seat/climbing log for the children but its main function was to create a habitat for living things – fungi, mosses and all sorts of 'minibeasts'. One part of the grassed area was deliberately not mown and allowed to grow wild. Here the children could find weeds growing, insects and other animals living in and on the weeds, leading to a wide variety of observational study.

The types of plant and animal to be found in the wild patch were found to be more numerous than they had thought and in recording them it was soon noticed that some plants were favoured by certain creatures for food. This led to investigatory work in the classroom. The creatures were brought in and their food preferences studied. To plan such an enquiry requires a number of decisions. Does it matter if the amounts of different foods are not the same? How many creatures should be used in the experiment? Where should they be put? How do we make sure that they know which plants are there? Problems of these kinds led to a variety of different approaches and decisions as to which was best. Do you give your caterpillar ten minutes on the nettles, then on the dock leaves, then on the rose-bay and see what he eats, or leave him to find out what is on the menu for himself? Whatever their methods all the children obtained some worthwhile result and were surprised to find the creatures able to recognize the difference between plants. *Just like us – I can't stand cabbage* said one. But behind the flippant remark may have been the recognition of some similarity, as well as of differences, among living things.

These descriptions provide some answer to the first question posed at the beginning of this chapter. For the answer to the second question – what do the children learn from these activities? – it is necessary to go further than description. Activities of the kind that have been described have to be examined and justified – as should all school activities – as being of educational value.

It is evident that the children were enjoying their work. But was their activity more than apparent 'play' with plasticine, blowing up balloons, looking for creepy-crawlies, throwing parachutes and paper aeroplanes? These activities are not unlike the ones that boys and girls spend their time on out of school, so why use school time for it? We must have a good answer to this question if science is to deserve and preserve its place in education at all stages.

Clearly it would be wrong to claim that from these particular activities the children involved will understand certain ideas, grasp certain skills or develop scientific attitudes. Ideas, skills and attitudes take time to develop and are formed gradually through the cumulative effect of many experiences such as these. So the question should be rephrased as: What do children learn from these activities and others like them which they encounter as a regular part of their school experience?

In the first class the experiences mentioned (and these were only a sample) would lead the children to learn some specific facts about their environment, some more general ideas, some skills relating to enquiry and some attitudes necessary for carrying out enquiries systematically and self-critically.

Specific facts about:
- the expansion of liquids when heated
- the points of the compass on the weather vane
- the size and variation in temperature during the day
- the amount of rain which falls at different times of the year
- the concept of 'prevailing wind' and its direction in their area

General ideas about:
- measurement by various different instruments
- the uncertainty in predicting the weather
- the relationship between different weather features
- volume and its relation to linear dimensions
- speed and its measurement

Enquiry skills of:
- making careful observations
- using a thermometer
- reading scales on various measuring instruments
- looking for patterns and relationships in data
- drawing graphs, filling in tables and charts
- proposing hypotheses
- finding ways to test hypotheses

Attitudes of:
- seeking and using evidence for making judgements
- patience and perseverance in collecting data systematically
- co-operation with others

In the second class the list is the same in some respects and different in others, particularly the specific facts.

Specific facts about:
- how a ship is launched
- which materials float or sink
- what shapes enable objects to float
- the tides
- some sources of pollution

General ideas about:
- pollution and its effect on living things
- volume and its relation to linear dimensions
- how the shape of an object influences its floating

Inquiry skills of:
- investigating variables which may affect investigations
- controlling variables in making comparisons
- planning 'fair' tests
- observing similarities and differences
- manipulating materials and equipment

Attitudes of:
- willingness to be critical of their procedures
- questioning
- perseverance

If we continued with this analysis for the third and fourth classes we would find areas of overlap between what is being achieved through different activities, showing that the achievement is built up through the combined influence of several activities. Also, overall, these examples would touch upon a considerable range of ideas, skills and attitudes, and it is not difficult to see that over one year or several years such activities could cover many of the goals of primary science work.

There is probably no need to continue listing items for the third and fourth classes, since clearly there *is* plenty that is being learned. The value of what is being learned lies partly in its immediate usefulness to each child in helping him to understand the world around and partly in the power it gives him to find out more through his own enquiry. These are the content-based and process-based goals of science whose nature and inter-relation require some further attention.

Content and process goals in primary science

There are two main components of learning in science. These are *a* the knowledge of facts and grasp of concepts (general ideas), and *b* the skills and attitudes relating to enquiry and finding out. Sometimes the impression has been given that these are separate and indeed that one is more important than the other. Fashions in education have sometimes led to an emphasis on one or the other; the 'object lesson' treated science as if it were only *a*, whilst the 'discovery approach' treated it as if it were only *b*. As we see things at the present time the two components seem inseparable, but this has not always been the case (see Harlen, *School Science Review*, 1978, and *Education 3 to 13*, 1980). There must be some content studied when skills and attitudes are encouraged and hence some facts and concepts will be learned. The existing facts and concepts known to a child will be the basis from which new experience will be understood. Enquiry, using the skills of observation and interpretation and the attitudes which enable these skills to be applied, will lead to further facts and concepts being revealed, which may necessitate modification of existing ideas. So as the process of learning proceeds, enquiry leads to the extension or confirmation of ideas, and the application of existing ideas throws up problems for further enquiry. A child's ideas are gradually changed and developed as his skills become refined and more powerful.

Hence it seems that ideas and concepts are inevitably developed when enquiry skills are used and, conversely, ideas and concepts are built up *with understanding* when learned through enquiry. The second part of this statement deserves emphasis and some explanation. It may seem perfectly possible to teach about the facts and general ideas which form the content of science by *telling* pupils about them. In parts of the world it is still not uncommon to see young children taught 'science' in the same way as they are told a story – the teacher talking and the pupils listening, periodically chanting back to the teacher words learned by rote with no understanding. It is quite evident that science, to these pupils, *is* like a story and means no more to them than that in relation to understanding their environment.

There are three chief reasons why teaching science content alone is unproductive. Firstly, the concepts which have to be grasped are complex and often abstract; children cannot understand them when presented in an elaborate and sophisticated form. Children must see concrete examples of ideas in application before they can have any idea of what they mean. (Imagine, for example, what sense the idea of floating would have for a child who had never seen water. This is unlikely in practice but the same confusion arises when children are told about air pressure without any way of visualizing what it means in practice). Secondly, presenting facts and principles in this way gives children no alternative but to memorize and accept what they are told (or not to, as the case may be). They have no way of questioning the so-called facts, or seeing for themselves that science principles are ways of explaining things which work as far as they can be tested but are always subject to question and revision. Without this basic idea being developed no science education as such can take place. Part of this argument is that the 'facts' of science change, a circumstance which does not matter if facts are viewed as explanations which can always be disproved, but is devastating if thought of as describing necessary reality.

The third point is one which is a matter of practicality. It has become a truism to say that the amount of fact there is to be known in science grows at an accelerating pace. Not all of this new knowledge, of course, has relevance for school science, but neither is all of it irrelevant. We only have to think of the change in the complexity of technology in the home and in children's toys over the past twenty years to realize the impossibility of conveying all the facts and principles which impinge on daily life. If science education results in children knowing *how* to learn and find out, being questioning and open-minded, it will do more for the future lives of these children than if it teaches them to memorize facts which become increasingly inadequate in explaining the world around.

Early scientific development

There is considerable consensus in the curriculum materials currently in use as to the range and nature of the aims or goals of science at the primary and middle school stage. Probably the most detailed list is provided by the *Objectives for Children Learning Science* of Science 5/13, (Macdonald Education, 1972). This contains a list of some 170 statements grouped under eight broad aims.
1 Developing interests, attitudes and aesthetic awareness
2 Observing, exploring and ordering observations
3 Developing basic concepts and logical thinking
4 Posing questions and devising experiments or investigations to answer them
5 Acquiring knowledge and learning skills
6 Communicating
7 Appreciating patterns and relationships
8 Interpreting findings critically

Within each of these broad aims the objectives are related to three stages of development (linked to a Piagetian description of development). These broad aims can be related to skills, attitudes and concepts and knowledge as in **Figure 1**.

Figure 1

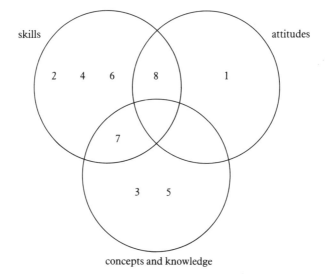

This shows that the majority of objectives relate to skills but that Science 5/13 does make some statements about ideas that should be learned, for example:

> Awareness of the properties which materials can have.
> Awareness of sources of heat, light and electricity.
> Knowledge of conditions which promote changes in living things and non-living materials.
> (Science 5/13, *Objectives for Children Learning Science*, 1972)

Since the Science 5/13 materials are for teachers and do not constitute a series of prescribed activities, these general statements of concepts are the appropriate way of indicating objectives. The ideas they convey can be arrived at through a variety of activities. An example of this approach in action is found in the work described at the beginning of this chapter – two quite different activities contributed to pupils' formation of the concept of volume and its relation to linear dimensions.

In other materials it is quite common to see more specific statements of the concepts pupils are intended to learn. For example the series called *Look!* (1981) consists of packs of workcards for pupils. The activities to be carried out are described and for each one there are stated 'general concepts' and 'card concepts'. For example:

> General concept: The surface of water can support objects. The surface 'sticks' to the objects.
> Card concepts: Detergent weakens the 'skin' on water. Some animals make use of the upper or lower surfaces of water.

The aims of the programme are stated as including enquiry skills, but these are implicit rather than explicit in the description of each activity. A possible drawback to this approach is that it is almost bound to emphasize concepts and knowledge goals rather than skill and attitude goals. This is because the former *can* be related to specific activities since they are tied to the *content* of the activity. The skills and attitudes, however, are achieved in the *way* the activity is carried out rather than in its content. Moreover they cannot be achieved through one or a few activities but only if a consistent approach is used over a whole series. Thus to state these objectives for each activity would become repetitive and lose any impact.

There is a dilemma here. One solution is not to give any statement of concept and knowledge goals at all for fear that attention to these will squeeze out thought about skills and attitudes. This leaves teachers without any guidance as to what ideas children should be gathering. Such guidance is necessary if activities are to be organized so that children can develop their skills of enquiry and at the same time encounter content which will build up useful concepts and knowledge. A more acceptable solution is to attempt to state all important goals at the same level of detail. This means that content-related statements must be worded so that they could apply in a range of activities just as do process-related ones. It avoids the problem of defining specific activities, a task which really must be left to teachers who know their children and the environment and are in a much better position to decide whether it is relevant for the children to learn about, say, the 'skin' on water, than are the authors of published materials.

The following lists are an attempt to give examples of these ideas in practice. They suggest the goals appropriate for younger (up to about nine years) and less mature older pupils, and for older and more mature pupils (up to thirteen or fourteen years). Everything proposed for the earlier phase is assumed to be equally relevant for the later phase but there are additions and changes of wording which reflect development of ideas and elaboration of skills.

Enquiry skills
For both earlier and later phases:
- observing and exploring to observe further
- raising questions
- proposing ways to answer questions through fair tests or comparisons
- finding patterns in observations
- classifying
- applying ideas in new situations
- finding out information using books, charts, etc.
- communicating information in various ways, by drawing, speaking, writing, etc.
- using simple measuring instruments

For the later phase – in addition:
- defining questions which can be answered by experiment
- proposing hypotheses
- identifying and controlling variables in carrying out investigations
- recording observations systematically
- using evidence critically and logically
- making measurements with appropriate accuracy
- communicating information in the most appropriate form

Scientific attitudes
For both earlier and later phases:
- curiosity
- willingness to put forward ideas
- co-operation
- perseverance
- open-mindedness
- care in handling living things

For the later phase – in addition:
- responsibility in carrying through an activity
- honesty in reporting results
- independence in thinking
- self-criticism

Concepts
For both earlier and later phases:
- about living things (needs, characteristics, variety, animal as including ourselves, plant, food, growth, development, life cycle, senses, health)
- about materials (variety and characteristics of the main groups)
- about change (in the sky, weather, living things, materials when put in water – sinking/floating, dissolving/not dissolving)
- about movement (causes, speed)
- about length
- about area
- about volume and capacity
- about mass
- about time
- about cause and effect

For the later phase – in addition:
- about basic life processes (reproduction, growth, feeding, respiration, sensitivity, movement and support, the variety in how these are carried out)
- about interdependence of living things
- about adaptation of living things
- about air (existence, use by living things, inclusion of water vapour, pollution)
- about water (occurrence, use by living things, solvent power, existence as solid, liquid and vapour, pollution)
- about the soil (composition, function, fertility)
- about the earth in the solar system (sun, moon, stars and planets)
- about forces and movement (sources and effects of force, stopping and starting movement, uniform and changing speed)
- about measurement (using arbitrary or agreed units, estimation, approximation)

Assessment – a challenge

Having looked at the nature of some typical science activities and at the variety and types of learning outcomes it becomes clear that assessment cannot be a simple matter. What is assessed may vary with the purpose of the assessment, as discussed in the next chapter, but it must always reflect the full range of goals. This presents quite a challenge which is not easily met in full but is well worth taking up.

We have seen that the learning which results from science activities is as much a product of *how* the children go about their work as it is of *what* activities they undertake. It follows that it certainly is not enough to record the activities the children have engaged upon. To assess the learning which has taken place it is necessary to find out about the ideas, the skills and the attitudes of the pupils and about the progress being made over time. Assessment has an essential role in education of any kind; unless we take assessment of science activities in the early years seriously we cannot be undertaking the science education of primary and middle school pupils seriously.

2 Purposes and methods of assessing children

Meanings

Sometimes the word *assessment* is used without there being a clear understanding of what it conveys. Some people use the word to mean the same thing as testing; some use it to mean the opposite of testing (as when performance on a course is graded by continuous assessment rather than on the results of a final examination); others consider it to imply 'formal' as opposed to 'informal' procedures. It will not be possible to clear up all these ambiguities by proposing a definition and expecting others to adopt it, but it is essential to make clear to readers the meaning of the word as used here.

Assessment is a process of gathering information, not the product of this process – though sometimes we speak of 'the assessment', meaning the grade or mark which is produced by assessment. In the process of assessment some attempt is made to apply some standard or criterion to the information. In some respects the process is rather the same as measurement except that the units used as standards are often implicit rather than explicit. When we measure the length of a piece of string in centimetres we compare its length with the length of the unit, the centimetre. When we find out how well a pupil can carry out a certain skill we implicitly compare what he does with some expectation or criterion. In both cases the result, the length in cm and the level of skill development, is something which emerges from the process of assessment and is distinct from the object being assessed. If we wish to convey information in these cases, but *not* to assess, we could do this by passing on the piece of string or some replica of the same length and, in the case of the child's performance, by making a record of his relevant actions or preserving the products of his work. Some record keeping is done like this, by collecting work and so being able to pass on the original work rather than a list of marks or comments on it. Thus an important characteristic of assessment is that it results in the replacement of the actual evidence by something which signifies a judgement of it.

However it is misleading to press the analogy between assessment and physical measurement too far. By doing so the impression is often given that the results of assessment have to be expressed quantitatively and this is not at all the case. Most assessment results are not quantitative, although, were there any point in it, they could be put in this form. The reason why they are *not* expressed as numbers is that it can be very difficult – and in many cases unwise – to find agreed units to use. In some situations the choice is open and it is a matter of preference of individual teachers. For example, some teachers write comments when marking a pupil's essay; these comments indicate their judgement of it – this part was *good*, this aspect was *careless*, the story showed *original ideas*, etc. These remarks imply that certain standards of what is *good*, *careful* and *original* had been applied. Other teachers might express the results of applying similar standards as 7/10 or B or 70 per cent. In other situations a qualitative result of assessment is the only obvious one. A teacher is often assessing pupils' understandings during discussions. She is not recording 7/10 or B in response to what she hears, but instead expresses her judgement by approving or disapproving words, by smiles or frowns, degrees of rejection or acceptance of the ideas put forward and other types of immediate response. At the same time there *are* situations where results of assessment are required in quantitative terms, especially if they are to be combined or compared with others.

In summary, then, assessment is a process through which information is provided about how some part of a child's behaviour compares with a level, an expectation or a standard. How these expectations or standards are selected and determined is considered later in this chapter. For the present we continue this general look at related issues by turning to the relationship between purposes, content and methods of assessment.

The 'why', 'what' and 'how' of assessment

The discussion in the previous chapter, about the meaning of science for younger pupils, gave some indications as to *what* might be assessed in this area. But before any detail can be settled about the content of assessment there is a prior question to be answered. This is the first of four key questions which have to be considered in planning and carrying out any assessment. **Figure 2** shows these questions arranged in the order in which they should be addressed.

Figure 2

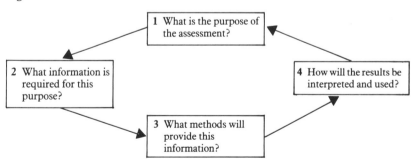

It is important to begin with question 1 since the answer will always – or should always – influence what is assessed and how it is assessed. For example if the purpose in assessing is to discover any difficulties pupils may be having, or to provide challenges which match their developing skills and ideas, then as wide a range of information as possible is required about what each pupil can do, regardless of where he stands in relation to others. On the other hand if the purpose is to summarize what has been achieved and to make comparisons between pupils or groups of pupils, then the range of information which can be used is restricted and it must be gathered using methods which make fair comparisons possible. Yet again, pupils are sometimes assessed for purposes which do not affect them as individuals – as in local or national surveys or in the context of research – and then the range and type of information will be different from that in the previous two cases. In all cases the information should relate to the nature of scientific development, or whatever area of experience is the focus of the assessment, but it will vary in breadth and detail. These variations, in turn, influence the assessment methods which can be used.

Various reasons 'why'

Lists of purposes for assessment can extend to considerable length; they are to be found in various general books on assessment (Deale, *Assessment and Testing in the Secondary School*, 1975; MacIntosh and Hale, *Assessment and the Secondary School Teacher*, 1976; Rowntree, *Assessing Students*, 1977). However the main points about different purposes can be brought out if they are grouped into major types. This is done in **Figure 3** which attempts to summarize the limitations and implications which relate to each type and have consequences for the *what* and *how* of their assessment. The five types of purpose are arranged in order of 'closeness' to the children they concern; closeness meaning the degree to which the results affect the children directly and immediately. Assessment carried out for diagnostic purposes is very close to children in this respect, whilst assessment for national monitoring is relatively remote, affecting the individual, if at all, only through policy decisions made sometime in the future.

Clearly the first three of these types of purpose are the ones which enter most into the work of teachers. The first is distinctly different from the second and third, since it is part of day to day work, intended to guide day-to-day action, rather than being an attempt to sum up achievement over a period of time. Chapters Three and Four discuss this day to day assessment in science, whilst Chapter Five deals with summary assessment in science. Little more is said about assessment for the last two purposes apart from a brief section at the end of Chapter Five on the national monitoring being carried out in science by the APU.

Figure 3 Types of purpose of assessment

Purposes	Implications for 'what' and 'how'	Limitations
Diagnosis: identifying problems, matching activities to development, planning by teacher, record keeping.	A wide range of information needed about progress in developing skills, attitudes, concepts. Methods must be capable of being applied frequently and repeatedly without interfering with normal work. No need for all pupils to be assessed in exactly the same manner since no comparisons are being made.	Information rich in detail but often arising from specific activities and not always generalizable.
Making comparisons: with earlier performance to show progress, to compare groups, to give information about individuals in relation to group or class.	Emphasis must reflect the basis upon which comparisons are made (which of necessity must be restricted). Methods are required to be such that pupils are all treated similarly and comparisons are fair.	May not be possible to find methods which allow for comparisons to be made *and* relate to all major types of goal. Often development of attitudes and skills is neglected for this reason.
Reporting achievements: to parents, pupils, other teachers interested in the progress of individual pupils.	Range of information should reflect all major goals but in summary rather than in minute detail. Methods must provide straightforward, readily grasped judgements which are reliable. The criteria used in the assessment must be made clear.	As above, existing methods may not cover all types of goal. Hence unless new methods are devised information tends to be about what is testable but not necessarily educationally important.
Research and curriculum evaluation: comparison of teaching approaches, materials, investigation of differences among pupils.	Information will reflect aspects which are the focus of the investigation, probably not the whole range of what is thought important. Methods must suit this focus without any in-built bias against particular approaches or pupils.	Information will be limited to what can be assessed in the short term, but this may not be a fair indication of progress towards longer-term goals.
Monitoring at national or LEA level: looking for differences related to school variables, for trends over time.	The assessment has to reflect a basic set of general goals which are valid for all schools even though their programmes vary. Methods must be able to be used with reliability in a large number of schools.	Results will be derived from assessment which is not tailored to the particular curriculum experienced by the pupils. Methods limited by their feasibility for large-scale use.

General points about 'how'

When considering methods of assessment there is a tendency to think straight away in terms of various kinds of test. Tests do indeed form a major set of ways of assessing but there are many other ways. For instance, within the range of meaning of assessment discussed above we can see that marking work, observing pupils' actions, listening to their ideas in discussion and other events in normal work can also provide information for assessment.

The word *test* covers quite a range of devices, but what they have in common and what makes them different from other methods is that they involve setting pupils tasks specifically designed so that the actions or responses can be assessed. In other methods the assessment is made of responses or actions which take place during normal learning activities. There are, of course, borderline cases, where a test exercise might be given as much for practice as for assessment, or where a test question might be incorporated as part of normal work, but, by and large, it is helpful to think of the distinction between tests and other methods as follows:

- assessment by tests: tasks set primarily so that assessment can be made
- assessment in non-test situations: tasks provided primarily as learning activities in which assessment can be made incidentally

Tests are sometimes sub-divided into various categories, such as *formal, informal, standardized, criterion-referenced, diagnostic* and so on. These labels are not always helpful since they refer to one feature of the methods only and are not an adequate guide when selecting the best method of assessment in a particular case. An alternative, which is now discussed, is to consider the features of assessment methods in general – of tests *and* of non-test situations – and then select or devise the method which has the combination of the most appropriate features for the information required for a certain purpose.

We begin by looking at four components which are common to all assessment methods. Any method of assessment must have:

- some way in which a problem or task is presented to the pupil
- some way in which the pupil responds
- some standards or criteria with which the response is compared
- some way of representing the result, as a mark, comment, or categorization

All methods have to have these components, but the way in which they vary is in the combination of different features of each. So let us consider the features for each component.

1 *Presentation of a problem or task*

The chief variations in ways of asking a question or presenting a task are as follows:

a On paper, using words, diagrams, numbers, pictures, etc. In this form the tasks can easily be presented to pupils individually.

b Using a medium other than paper – showing or demonstrating something, using actual materials or films or tape recordings. It would generally be impractical to do this for individual pupils, and so presentation would most likely be to a group or class.

c Providing practical equipment such as paints, wood, science or maths apparatus for pupils to manipulate.

d Normal work – this might include any of the above.

2 *Ways in which pupils can respond*
In acting or responding in the situation presented, the pupil might be involved in the following:
a Selecting from alternatives provided by simply ticking. This is most likely to be a response to a task presented on paper.
b Producing written answers – either long or short. This could be a response to any of the ways of presenting a task.
c Drawing, painting on paper.
d Constructing or manipulating apparatus and materials.
e Speech, gesture or action.

3 *Standards or criteria used in judging the response*
When the response is made the process of assessment involves comparing it with the standards or criteria which will result in it being able to be categorized. These standards can be derived in the following ways. Each of these could, in theory, be used in relation to any of the types of response in 2.
a Reference to what is the norm, or average, of the group to which a pupil belongs. In this case a pupil's mark, or comment from the teacher, will depend on how other pupils perform as well as on his own performance. In formal standardized tests the 'norm' is a certain mark, but in less formal methods of assessment a teacher may still be applying standards derived from experience of other pupils in deciding what is an acceptable performance.
b Reference to criteria which indicate that a certain level of understanding or development has been reached. In this case the result of the assessment will depend on how a pupil's performance matches up to the criterion, and is independent of how other pupils perform. The derivation of criteria, however, arises from what it is reasonable to expect of pupils of a particular age and experience. Thus criteria are related to norms in derivation though they are applied differently when being used.
c Reference to a pupil's previous performance or to the teacher's expectation of the pupil. This may be termed *pupil-referenced* and occurs when, for example, a teacher puts a favourable remark on a piece of work of one child even though it is of poorer quality than the work of another, of whom the teacher expects more.

4 *Ways of presenting the results of assessment*
a Putting a tick or cross, to indicate right or wrong. This is the simplest application of some criterion of what is right or wrong, or acceptable or unacceptable, for the pupils concerned. (Where this marking can be done without the marker exercising any judgement about borderline cases, the marking is described as *objective*). A glance at the types of responses in 2 shows that this way of marking is likely to be applicable only for 2a and possibly for 2b.

b Expressing the result in terms of a scale, such as marks out of 10 or A to E. Sometimes this will result in a few As, a few Es and most marks being B, C, or D if there is norm-referencing taking place. If criterion-referencing is being used there should be no limit on the proportions of different grades or marks.

c Using qualitative categories, usually in the form of a set of descriptive criteria which are applied on the spot (in the case of actions or speech) or to a written record or recorded response. Use of these may result in a comment, oral or written, or marks against items on a checklist.

Summarizing these features, it can be seen that the possibilities are:

Presentation of task	*Response of pupil*	*Basis for results*	*Representation of results*
1a On paper b Demonstration film, tape, etc c Practical situation d Normal work	2a Selection b Writing c Drawing, etc d Constructing e Speaking, action	3a Comparing with the standard of others b Comparing with criteria of performance c Comparing with pupil's previous performance	4a Tick or cross b Number or grade c Comment, or qualitative categorization

Different methods of assessment can be described, or created, by taking combinations of these features. For example, the standardized multi-choice objective test has the following features:

1a + 2a + 3a + 4b

whilst a teacher-made practical test in science might have the following features:

1c + 2d + 3c + 4c

An English essay could be an assessment with the following features:

1d + 2b + 3c + 4c/4b

In the latter two examples much would depend upon the purposes and so we can only suggest possibilities. Indeed the value of this analysis is precisely in the matching of methods to purposes and reference will be made to these features in discussing assessment for different purposes in the following chapters.

A simple example of how the features relate to purpose is the choice of the basis for producing a result. If the purpose of assessing work is for feedback to the pupil and to show whether progress is being made, then pupil-referencing (**3c**) is an acceptable option. However, if results are to be used to show where a pupil is in relation to others (as when class results are passed on to other teachers), then pupil-referencing is not acceptable and criterion-referencing (**3b**) would be preferable. A version of norm-referencing (**3a**) is often used in such circumstances to show the relative positions of pupils, but this is open to objections when applied over as small a number as in one class, or even one age group in a school. Moreover, it is arguable as to whether a certain distribution of marks should be imposed on the results relating to what has been learned (which should not be randomly distributed!). Not all possible combinations of these features can exist in actual assessment methods, of course; for instance there is unlikely to be a method of assessment combining an oral or action response by the pupil with a method of marking which does not make some demand on the judgement of the marker.

Examples of using the features

The analysis of assessment methods in terms of their features has two main applications in the matching of methods to information required. In the first of these it can be used to examine methods which are being, or may be, used in assessment to reveal aspects which make them unsuitable for what they are supposed to assess. The fact that a method involves a particular way of presentation of the task, or requires a response in a certain form, may make it incapable of giving information about some behaviour. The ability to manipulate equipment or skill in using measuring instruments (not just the knowledge of how to do these things) clearly requires an action response; if assessment is only by means of paper and pencil then something other than the manipulative skill is being assessed and this should be realized. Similarly, creative writing cannot be assessed by a method which involves selection from alternatives and cannot be marked right or wrong.

An example may help to show how the analysis in terms of features can be useful in assessing the adequacy of existing methods. A teacher of fourth year primary children wished to record, both for her own use and as a basis for informing parents and other teachers, information about the progress of her pupils over the whole range of their activities. She had a wide range of goals in mind for her pupils, but it is enough to consider just a sample of them: ability to observe, skill in manipulating numbers, development of responsibility, originality and the concept of life cycle.

The existing methods which the teacher used in compiling records consisted of marking work (marks out of 10), occasional home-made tests of general knowledge, arithmetic and spelling, and standardized reading and verbal reasoning tests given occasionally. The features of these methods are as follows:
- marking work: **1d + 2b + 3c + 4b**
- home-made tests: **1a + 2b + 3b + 4a**
- standardized tests: **1a + 2a + 3a + 4b**

Presentation on paper, as in the two kinds of test, severely restricts the possibility of assessing ability to observe, the expression of originality or evidence of responsibility. Thus, of the five examples, only the numerical skill and the concept of life cycle could be adequately included in the tests.

Work arising in day to day activities could, potentially, provide opportunities for all kinds of development to be expressed, assuming that the work was related, in practice, to all the goals. But in obtaining information, only the marking of written work was taken into account. This raises the question of how much of children's development in observation, responsibility, originality, even the concept of life cycle, would be reflected in their written work. A look at the workbooks of a few ten year olds will reveal that ability to express these things through their writing is rare. Hence the methods used by the teacher cast doubt on their appropriateness for assessing even these five items. To correct this state of things the teacher might examine again the features of assessment methods which would give her the kinds of information she required.

This takes us to the second way in which the features may be of use, which is in suggesting methods suitable for particular purposes. An example comes from a First School which was faced with the situation of having no records which indicated the emphasis the staff placed upon the personal development of the children. They wished to expand their records and gather information to serve two purposes: to indicate to parents, managers and the Middle School to which their children transferred, that the school worked seriously to achieve goals in addition to basic numeracy and literacy; and to enable the staff to see how effectively they were succeeding in the development of personal and social attributes in the children. Between them the teachers drew up a list of information needed to serve these purposes. It included enthusiasm, originality, security, responsibility, co-operation with teachers, co-operation with other children, self-criticism.

In the next step it was necessary to look at the suitability of the features of each component for assessing attributes of these kinds. The results can be summarized as follows:

- *presentation of task:* **1a** not feasible for First School pupils even if written methods of assessing the desired attributes existed. **1b** and **1c** would require a great deal of preparation. Only **1d** provided a possible feature – normal school activities with no special situations created.
- *response of pupil:* Once the agreement had been reached about using normal work for the assessment, then **2a** became irrelevant. **2b** could also be excluded because of the age of the pupils. Some information might come from examining the children's drawings or paintings and their models (**2c** and **2d**) but most would be found in their actions and speech (**2e**).
- *basis for results:* **3a** was felt not to be appropriate; instead it was considered that certain types of behaviour should be recorded where they were found, regardless of the general level of such behaviours. Thus **3b** was the feature selected.
- *representation of results:* Given that criteria would be applied to the pupils' responses and actions the best choice might well be thought to be recording in terms or categories related to these criteria (**4c**).

Initially, however, the decision was to grade behaviour along a five-point scale for each attribute (**4b**).

From this analysis the following set of features emerged as those of the assessment method required: **1d + 2e + (2c/2d) + 3b + 4b**.

In practice what these decisions meant was that teachers were expected to record those observations of pupils in their normal work which were relevant to the attributes being assessed. An assessment sheet was drawn up which enabled observations to be recorded for each child, as in **Figure 4**.

Figure 4

Unenthusiastic					Enthusiastic
Lacks originality					Original
Insecure					Confident

and so on

There were several problems revealed when this was put into practice. Some tidying was needed to cut out overlap between attributes and to reduce the overall list to manageable proportions, but beyond this there was a basic worry that teachers were doing no more than record their highly selective observations and subjective judgements about the pupils. The problem was that there were no agreed criteria being used in the assessment; it had been a mistake to assume that a system of grades would be a satisfactory way of presenting the results. What had happened was that, whilst the information might be valid, its reliability was unacceptably low due to the decision about the representation feature. A decision was taken to change this feature from **4b** to **4c**. Categories of behaviour describing different points in development of each attribute had then to be defined. Of course this raised a whole set of new questions. But we can leave this example at this point since these are among the questions taken up in the next chapters in relation to assessment in science.

3 Day to day assessment of scientific development – 1

Introduction

The focus of this chapter is assessment which is carried out to help day to day work. One of the main characteristics of this type of assessment is that it provides immediate feedback so that action can be taken based on information about children's progress. It is in this characteristic that it contrasts with assessment which takes place at the end of a period of work, when it is too late to use the information to change activities or to provide help which may have been needed. A second characteristic is that it is detailed and covers a wide range of goals, relating to skill, attitude and concept development. This detail enables it to fulfil a diagnostic function, remembering that diagnosis is not only relevant when there are problems. The meaning of diagnosis is very close to 'knowing about' the object in question, a process which is appropriate at any time, whether or not it is thought that remedial action is required.

The role of assessment in matching

The importance of this kind of assessment in some areas of the curriculum is so obvious and widely accepted that it scarcely needs mentioning. In the teaching of reading, for instance, the feedback to the teacher from hearing a child read is as important a part of this activity as the practice and challenge it gives to the pupil. The teacher generally uses this feedback immediately in decisions about how to help a child further with reading. In science activities, however, whilst there is the same kind of role for feedback to play, the process takes place much less frequently. The reasons are probably many and various, relating both the complex nature of the things lumped together in the phrase 'scientific development', and to the background knowledge and training of teachers. Whatever the reasons, the result is now well-known, that science activities are less well matched to pupils than activities in any other part of the curriculum, as shown in the evidence of the HMI survey of primary schools (DES, *Primary Education in England*, 1978). Given this situation it is not inappropriate to take a little time considering the role of assessment in matching.

The range of activities and materials with which pupils come into contact in the classroom depends essentially on decisions made by the teacher. The teacher is not, of course, acting in isolation, being on the one hand subject to constraints and on the other able to benefit from help and guidance, which originate in the school and outside it. But the teacher has ultimately to take the decisions about activities in the classroom. These may be about such things as whether, for instance, to let David and John go ahead with their investigation of the smell of coloured sweets, or to encourage June and Stephen to wire some traffic lights for their model street, or whether to start a class project on colour following enthusiasm about a tie-dying session. These things have to be decided in the end by the teacher. This does not preclude pupils making suggestions for their own investigations, for even when suggested these cannot take place without the agreement and planning of the teacher.

The point of relevance here is whether the teacher makes these decisions with or without regard to the match of the activities to the pupils. If the latter, then there is less chance of the activities contributing as much as they could to the pupils' development. There is a great deal packed into the word *matching* here and perhaps we should unpack it a little. Some take matching to mean giving pupils activities for which they already have the necessary ideas and skills and in which they are therefore likely to succeed. The value of the activities would then be in the practice and the satisfaction of success they bring. However, such satisfaction is often short-lived and soon gives way to boredom; furthermore, it means that little, if any, progress is made. Alternatively, matching can be conceived as dynamic rather than static, as a process of enabling steps to be taken in developing some ideas or skills, where the aim is change and progress. The keynote would be the conviction that the child engaged on an activity is 'getting something out of it'. It is in this sense that we are using *matching* here. When conceived in this way it becomes clear that for matching to be contrived, it is essential to know something about where the children are in their ideas or skills. Thus assessment of these things becomes a crucial part in this process and is also related to knowing what are the potential next steps in development to which a matching activity should be leading.

So it may be, in the example of David and John, that their investigation of the smell of coloured sweets would be little more than a repetition of a previous activity unless the teacher discusses with them the additional variables they might control, leading them to a more sophisticated experiment which would be a greater challenge to them. In the case of June and Stephen, it could be that they would need to investigate simple circuits before embarking on the design of their traffic lights, otherwise they would be simply following instructions without understanding. In these cases consideration of the existing ideas and experience of the children could lead to their gaining more from their activities than would otherwise have been the case.

It is worth noting at this point that the information the teacher needs for these decisions has to be at her command at almost any time. She will need to use it not only in decisions before activities begin but at any time in helping or extending activities. If she does not already have the information she needs at these times it will be too late to start looking for it when decisions need to be taken without delay. Hence the assessment has to be going on all the time, so that gathering information about the children is part of the process of teaching and not a separate matter where teaching stops and assessment takes place for a while. This will be a significant feature to keep in mind when we come to consider methods of gathering information. Before doing this, however, remembering the order of questions in **Figure 2**, we must turn to the question of what information is required.

Information required as feedback

In attempting to match learning experiences to the points reached by children in their scientific development it is clear that information may be needed about all the goals of science education which were mentioned in Chapter One. This list *could* be very long indeed, making the task seem quite formidable. For example, in the comprehensive list of objectives proposed by Science 5/13 (page 7), the objectives are arranged in three stages of development, so for any one stage a smaller number would be relevant. Nonetheless there would still be too many for information about each to be gathered. Furthermore, Science 5/13 statements were not devised to be used as a basis for assessment; there is overlap between them and this matters less in guidelines for devising and deciding activities than in a framework for assessment.

For the purposes of assessment the list should:
- be as short as is compatible with being complete
- contain items which are clear and unambiguous in meaning
- contain items which overlap as little as possible, though at the same time leaving no significant gaps

One example of a list intended to serve as a framework for gathering information about pupils is one which was produced through discussion among teachers in several groups as part of the Progress in Learning Science Project. This list (**Figure 5**), was proposed for upper junior and middle school years; a slightly different list was proposed for younger and less mature pupils.

Figure 5

Skills of:	*Concepts of:*	*Attitudes of:*
observing	cause and effect	curiosity
proposing enquiries	measurement	originality
experimenting/investigating	volume	willingness to co-operate
communicating verbally	force	perseverance
communicating non-verbally	energy	open-mindedness
finding patterns in observations	change	self-criticism
critical reasoning	interdependence of living things	responsibility
applying learning	adaptation of living things	independence in thinking

Single word titles cannot make clear the meaning of each one and some expansion was offered, as in these examples:
- *observing:* noticing all kinds of details or changes, using several senses where possible, as shown by comments, records made, or actions taken at the time or later; distinguishing between observations relevant to a particular problem and those which are not relevant
- *concept of change:* appreciating that change is not spontaneous but always results from some interaction even though the mechanism may not be known; being aware that there are different kinds of change and of the circumstances in which these take place
- *open-mindedness:* being generally prepared to listen to other points of view; accepting ideas which are new if the evidence is convincing after consideration

(Match and Mismatch: Raising Questions, 1977)

Among the twenty-four items in this list, covering about equal numbers of enquiry skills, concepts and attitudes, there are several which overlap considerably with other areas of the curriculum. Compare some of the above with the list of objectives of a combined history, geography and social science project, shown in **Figure 6**.

The overlap in skills and attitudes (or personal qualities) is so striking that if the same teacher or team were concerned with both areas of the curriculum there would be little point in having two lists. There could be a general list – which might well be extended to cover mathematics as well – and then items more specific to the concepts developed within each subject could be encompassed in very short lists. However this would only be satisfactory *if* the general items were being assessed in other areas *and* the teacher had access to that information when helping the children's work in science. If this were not the case – if a different teacher was concerned with science, for instance – then it would be inadvisable to reduce the list, since the purpose is not to ensure that progress towards these goals is somehow assessed and recorded, but to have the information to hand to use in daily work.

In drawing up any list of goals in a form suitable for indicating the information needed for day to day assessment it has to be recognized that what is included ultimately depends upon judgement as to what is worthwhile. The result will reflect the views of those who construct the list about the subject in question and about education in general. This is completely unavoidable since all such issues in education have to be settled by discussion and consensus. Straughan and Wrigley have pointed out: 'A vast amount of work still remains to be done in identifying the wide range of knowledge, skills, attitudes and qualities which are held to be of value in education today within all the main subject areas of the curriculum. Obviously there will be less consensus in some areas than others, and controversy is bound to arise generally over the ranking of priorities. But there is no reason why the

Figure 6

	Skills		Personal Qualities
Intellectual	Social	Physical	Interests, Attitudes, Values
1 The ability to find information from a variety of sources, in a variety of ways. 2 The ability to communicate findings through an appropriate medium. 3 The ability to interpret pictures, charts, graphs, maps, etc. 4 The ability to evaluate information. 5 The ability to organize information through concepts and generalizations. 6 The ability to formulate and test hypotheses and generalizations.	1 The ability to participate within small groups. 2 An awareness of significant groups within the community and the wider society. 3 A developing understanding of how individuals relate to such groups. 4 A willingness to consider participating constructively in the activities associated with these groups. 5 The ability to exercise empathy (i.e. the capacity to imagine accurately what it might be like to be someone else).	1 The ability to manipulate equipment. 2 The ability to manipulate equipment to find and communicate information. 3 The ability to explore the expressive powers of the human body to communicate ideas and feelings. 4 The ability to plan and execute expressive activities to communicate ideas and feelings.	1 The fostering of curiosity through the encouragement of questions. 2 The fostering of a wariness of overcommitment to one framework of explanation and the possible distortion of facts and the omission of evidence. 3 The fostering of a willingness to explore personal attitudes and values to relate these to other people's. 4 The encouragement of an openness to the possibility of change in attitudes and values. 5 The encouragement of worthwhile and developing interests in human affairs.

(Blyth et al, *Place, Time and Society 8-13*, 1975)

exercise should not be attempted and a rational debate initiated.' (*Values and Evaluation in Education*, 1980). There is no 'correct' list of goals and hence of information which is needed about progress. What is important in any particular school is that a list should exist which has the support and agreement of the teachers. Whether there is one general list or one for each subject area will depend upon the organization of the school, but if there are separate ones they should be compatible with one another so that a consistent educational policy is pursued in the various parts of the children's experience.

Possible methods of assessment

When it comes to considering how to gather the information needed to help with day to day work, the nature and breadth of what is required presents a formidable problem. To meet its purpose any method of assessment should have the following ideal properties. It should be:
a flexible and usable repeatedly in a number of different situations
b able to give immediate results, without delay for marking and processing marks
c used without consuming a disproportionate amount of teaching and learning time – the information would defeat its purpose if gathering it were to be so demanding of teachers' preparation time and pupils' lesson time that it was a serious interruption of normal work
d able to cover the range of attitudes, skills and concepts relevant to the goals of the work – this follows from the argument that feedback is essential for the provision of effective learning opportunities, so it is necessary to know where children are in relation to all genuinely held goals
e able to give information of acceptable reliability. Reliability in this context means the extent to which the same result would be obtained were the same assessment to be repeated

In relation to the last point it is worth adding that for the purpose we have in mind in this chapter, the reliability of any one item of information need not be as high as is required if the results are to be used for a different purpose, such as to grade or rank a pupil against others. In the present case, the constant repetition of the assessment allows for correction of earlier results which may have been made in error. Nevertheless there is no excuse for not making assessment as reliable as it can be. It is not only a waste of time to gather information which is known to be highly inaccurate, but it would also be foolish to use it in taking decisions. So it is always worthwhile to attempt to obtain results which are as reliable as necessary for their purpose.

The list of ideal properties can now be considered in relation to the

possible features of assessment methods discussed in Chapter Two and summarized on page 18.
- *presentation of task:* Preparing written tests or demonstrations would be too time-consuming for regular use day by day. Besides, written tests would not be suitable for assessing practical skills or attitudes such as open-mindedness or co-operation. The main possibilities are **1d** (normal work) or **1c** (special practical situations).
- *response of pupil:* The nature of the range of goals to be covered suggests that a range of communication channels should be used. Only **2a** (selection from alternative responses) can be eliminated since it is tied to written tests.
- *basis for results:* As the purpose is essentially to help each child the information will not be made more useful by comparison with averages or the standards of others. It has to be interpreted in terms of what a pupil can do and so it is most useful to have criteria of performance indicating certain points in development. To interpret results in terms of progress, pupil-referencing will also be important.
- *representation of results:* Information for this purpose has to be rich in detail, thus ticks and crosses, numbers or grades, could not communicate in themselves this quality. The best alternative seems to be **4c** (comment or qualitative categorization). It is not always necessary for a written record to be made, for most often the results will be used in guiding short-term action.

The two methods emerging as most likely candidates for day to day assessment are ones which use either *a* normal work or *b* specially set up practical situations, where pupils' responses in writing, drawing, construction, action or speech would be taken into account. The responses would be compared with criteria indicating development and the resulting assessment used in making decisions about appropriate activities and approaches.

The main choice here is among *a*, *b*, or a combination of these. The special situation need not be all that different from normal work – in fact, as far as the pupils are concerned, they could be indistinguishable – and therefore need not be considered as reducing learning time. The kind of special situation in mind here might be, for example, a practical investigation where the pupils are bound to encounter variables, some of which should be kept the same whilst others are changed in a controlled manner. The teacher would then *know* that the opportunity to observe the pupils' reactions to this situation would arise and that she would be able to watch for it. In normal work the teacher might miss opportunities to assess such things since they might not be planned to occur. A counter-argument, however, is that if normal work allows opportunities for the skills and ideas to be developed then it also provides opportunities for them to be assessed. Further advantages and disadvantages of these two approaches can be summarized as follows, **Figure 7**.

Figure 7

	Advantages	Disadvantages
Normal work	Information can be gathered from the whole range of activities. Pupils' responses are more likely to be a valid indication of what they can do. No extra preparation is required.	Key events may be easily missed. Opportunities for observing certain skills/application of ideas may not arise unless specifically planned. Pupils' reactions may be influenced by the group in which they work.
Special activities	Can be sure that certain skills, ideas, etc will be required. Teacher can be prepared and ready to note pupils' reactions.	May seem artificial to pupil and so not engage his complete effort. Take time to prepare. Same activity cannot be used too often. Limited time of observation means that many such activities have to be devised and prepared.

Many teachers find that the greatest obstacle to using observation in normal work is that it is not easy to know what to observe. Putting pupils in specially set up situations enables certain skills to be observed more carefully and is a way of training the eye and ear to know what information to pick up. Thus a practical compromise is to progress from:
   special situations ⟶ normal work ⟶ normal work only
                       + special
                       situations

Such a transition was described in relation to the use of criteria drawn up by the Progress in Learning Science Project. One of the members of the team worked with teachers in her school who had had difficulty gathering information from children's normal work. As one put it: 'I don't know what I'm looking for and I might put something down which is not useful.' (*Match and Mismatch: Raising Questions*, 1977). So they gathered together with their coffee in a quiet corner of the classroom and watched whilst a child was asked to sort out a collection of shapes of different materials. The teacher talked to the child and drew out some of his ideas about area, classification, type of material, etc. Afterwards they discussed the actions and responses of the child in relation to the criteria. The other teachers then did the same with other children in the next few days, and it was not long before they reported seeing children do things which they had never noticed before in the course of their normal work. As these were class teachers, working with the same children all day, they soon found that they didn't need to use special activities, for they could find information about the children's progress from all the kinds of activities taking place in any case.

4 Day to day assessment of scientific development – 2 (Examples)

Assessing by observation in normal activities – an example

As an example of 'observation of normal work' in action, consider the following series of events, concerning Kevin, Andy and Joss. The numbers in brackets in this account relate to points commented upon afterwards.

It started with a fairly popular activity. The class topic for science was 'air', and Kevin, Andy and Joss were blowing through a tube to displace water from an upturned jar in a sink to see how much air they could each produce from one 'blow'.(**1**) At first they compared the amounts they blew out by measuring the distance the water level in the jar dropped. They argued that this was fair because the jar was straight-sided.(**2**) But then Andy, whose result had been the lowest and also had been the last to blow, thought that it might have been easier to displace water from a full jar than to start with it half-full of water. So he suggested that they should always start with the jar full of water.(**3**) But it was difficult to get it completely full, given the small size of their sink, so they decided it was fair for everyone to start with water at the same level, which they marked on the jar. Each time they began they adjusted the starting position of the water level to this point by careful blowing.(**4**)

In these next trials Andy was determined to get a better result and took an enormous breath and blew out until his face was red. (The teacher had warned them to be careful not to suck back water into their mouths, so one person made sure the tube was taken out of the blower's mouth as soon as bubbles stopped coming). Andy's result certainly was much bigger this time. He said this showed that it *did* matter about where the water level was to begin with. Kevin's result was also different the second time but this didn't seem to make Andy change his mind about the cause of the difference in his case.(**5**) Joss was keeping a record of the results and didn't know what to write down, as when every time someone tried the result was different.(**6**)

A discussion with the teacher at this point resulted in the suggestion that everyone should be allowed three chances and the largest result would be taken.(**7**) A table was drawn up and results were entered into it. Kevin's result was the biggest in the end and Andy's next. Joss said he thought Andy's would have been most because Andy was the 'biggest' of the three. *Depends what you mean by biggest* said Kevin, *Andy's taller than me, but I bet my chest's bigger than his*.(**8**) It wasn't long before they had added two columns to their table, for height and chest measurement. The result was not very clear at all.

When they came to report on this work to the rest of the class they explained why they had measured their height and chests. This started a discussion which soon ended in the class dividing into supporters of the 'height' theory or the 'chest size' theory.(**9**) The teacher asked *How could we find which is right, if either of them is?* She had to prompt to get the suggestion that more results from different people would help. But the idea was taken up eagerly and plans made for other members of the class to do their blowing (each with a clean piece of tubing) and be measured.

It took some time to gather all the data, and the teacher noticed that the three investigators often cut corners in their method, not always adjusting the initial water level carefully and not always giving everyone the 'best of three' as they had done for themselves.(**10**)

Making sense of the results was even more of a problem when all the class had taken part. Kevin's idea was to see if the person with the biggest blow was also the tallest or had the biggest chest. This didn't work because the biggest blow was from someone who was neither the tallest nor largest round the chest. It was difficult to see any pattern because none of the results was sequentially ordered.(**11**) The teacher set them the task of putting the results in order, first of all from the tallest to the smallest, *unless you can think of an easier way*. She gave them some squared paper and showed them how to plot out the results at the same time as putting them in order.(**12**)

Now we consider some of the observations made by the teacher during these activities which helped in the assessment of these children's progress. Some of these were *direct* observations, made by watching what was going on, some were *indirect*, made by discussing with the pupils what they had done and sometimes asking them to 'replay', so that the teacher could see how they had got to the point they had reached. Such recapitulation is essential for the teacher and a useful form of reporting for the pupils.

1 It was surprising that they paid no particular attention to ensuring that only one blow was taken. Often pupils insist on holding noses in this activity to make sure no second breath is taken. Had they missed this variable, or decided no special steps were necessary to control it? She asked them about this later and was told that they thought it was enough to tell people to use only one breath. A nice answer but not one which suggested that they had appreciated the need for control of this important variable.
2 A possible variable had been checked here and seen to be not influencing the result. It was interesting, too, that they accepted the *length* of air displacing the water as a measure of the amount. At one point a suggestion of calibrating the jar in units of volume was dismissed because it was unnecessary.
3 No-one challenged this suggestion, though it could have been tested out.

4 This was time-consuming but it did not occur to them to find out whether or not it really was necessary, i.e. whether the starting level made any difference.
5 Further evidence that Andy, at least, was jumping to conclusions without checking other possibilities.
6 It was quite late on in the investigation that Joss began to write anything down and he only did this when accused of not remembering an earlier result correctly. What he did write was disorganized and quite a long way from the table which the teacher eventually suggested.
7 The 'best of three' approach was chosen after an unsuccessful attempt by the teacher to suggest that an average should be taken. The boys did not seem to be able to recognize different results as being inevitable. To them they were 'wrong' because not so good as the best. Perhaps this situation, where their personal performance was under test, was not the most appropriate one for discussing error in measurements. The teacher decided to await a better opportunity to introduce this idea.
8 Although Kevin had challenged Joss to clarify what was meant by 'biggest', the number of possible meanings they had discussed was very limited. Another relevant meaning which they could have followed up was weight. Again it was necessary to ask if they had thought about this and dismissed it, or not considered it.
9 The rest of the class seemed to accept the two hypotheses offered, rather than challenge that other factors could be more important. However the supporters did back up with reasons for supporting one side or the other, though many reasons could not bear critical appraisal (e.g. *If he's taller there must be more air in him, mustn't there?*)
10 Assuming that they were convinced in the beginning that the steps to control variables and make the comparisons 'fair' were necessary, then their failure to carry these out consistently was more a matter of attitude than of understanding. Though they may have thought the controls to be desirable rather than essential.
11 The children were stuck for any way of looking for patterns in their data, after the failure of Kevin's first idea. The teacher realized that they needed help and practice in finding patterns in much simpler data. In the present case the data were too complex for them to make much headway on their own. She gave them a great deal of help so that they had the satisfaction of achieving a result after so much effort.
12 They picked up the idea of graphing results roughly to show up patterns and had no difficulty in plotting their results.

So, by watching, by questioning and by listening the teacher obtained information about these children concerning their:
- awareness of the variables operating (**1, 2**)
- ability to control variables (**10**)

- understanding of volume (**2**)
- willingness to be critical of their procedures (**4**)
- willingness to question or accept ideas proposed (**3, 8, 9**)
- perseverance (throughout)
- ability to keep systematic records (**6, 12**)
- appreciation of the need to repeat measurements (**7**)
- critical reasoning (**9**)
- ways of carrying out investigations (**10**)
- ability to see patterns in findings (**11**)

To assess these abilities and attitudes the information gathered from this or other activities has to be interpreted in terms of the children's progress. The question of the criteria to be used for this purpose is taken up later in this chapter (page 36). For the moment we turn to some examples of gathering information by observation in special situations.

Assessing by observation in special situations – examples

One of the American projects (Science Curriculum Improvement Study) provides a series of highly structured units, for each of which there is an 'Evaluation Supplement'. These suggest special activities which may be used with a whole class or a group. In the case of a unit on 'Life cycles', for instance, the teacher is instructed to prepare for the assessment by collecting together a range of living things, such as a pot plant, onion, tadpole, worm or any plant or animal brought in by the children. The instructions continue:

> Gather four to six children at a table and show them the collection of organisms. After the children have had time to examine the collection, separate the organisms into two groups – plants and animals. You might ask:
> *How do you think I sorted the organisms?*
> For children who do not respond, ask:
> *How are these two groups different?* or
> *How are all of the organisms in one group alike?*
> After each child has had an opportunity to respond, direct several convergent questions similar to those below to less responsive children, to those who appeared to rely on the ideas of others, and to those whose responses did not include differences between plants and animals.
> *Do these organisms (either group) move by themselves?*
> *Do these organisms (either group) have legs (roots, flowers)?*
> *Do these organisms (either group) have seeds (eggs)?*
> If no-one has suggested it, ask:
> *Are these plants (or animals)?*
>
> *Criteria*
> Plan to record the outcomes immediately after the discussion. Use the

criteria below to help you recall each child's participation in the discussion.
1. Did he generalize by naming at least two ways in which plants and animals differ (or name ways in which all plants or all animals are similar)? Appropriate responses might include references to structure (*these all have mouths and those don't; these all have leaves sometimes*), means of reproduction (*these have seeds and those have babies or eggs*), movement, and means of acquiring nutrition.
2. Did he, when asked convergent questions, recognize two or more properties that distinguish between plants and animals?

(SCIS, *Life Cycles: Evaluation Supplement*, 1973)

A second example is provided by the *Activities for Assessing Classification Skills*, by Gall-Choppin, 1979. These are materials which can be used by individual pupils, groups or a whole class. They:

'... aim to help teachers find out what stage of development has been reached by the child in performing classifications. It is expected that the knowledge gained about individual pupils can be used as an aid to guide the children's future work.'

The materials are in the form of worksheets on which are printed pictures of shapes or objects to be grouped and classified. Each child has one sheet of pictures and an answer sheet. In some activities the child is instructed to cut out the pictures, sort them out into groups and stick them on to the answer sheet. The child has to name the groups and give some explanation for the grouping. In other activities children are asked to use suggested groupings and to say how many objects of certain kinds there are in each group. Detailed instructions for the teacher indicate what help to give to children who ask for it and how to assess the results.

'The child's responses do not yield a standardized score but are judged against a set of criteria for success. Interpretation is based on what the child can or cannot do within each activity, and is primarily aimed at assisting in diagnosis of difficulties.' (ibid)

Whilst the first two examples rely on what the children do or say as evidence of the ideas and skills they have, this third example shows how mental and physical skills could be assessed by observing how they carry out an activity. It would be necessary to have one child, or a group working at a time on an activity specially devised to require, for example, the recognition and control of variables or the making and interpretation of observations.

The content would need to be chosen to fit in with the topic of current activities and so would be best decided by the teacher. Suppose the activities were centred on sound and musical instruments. One activity could be set up for an individual or a group to investigate whether the thickness of a rubber band made any difference to the sound it makes when stretched round a box and plucked, (**Figure 8**).

The equipment provided would be: a suitable open box (e.g. sandwich box); several rubber bands of different thickness but the same length;

Figure 8

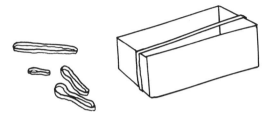

longer and shorter rubber bands of the same thickness.
The pupils could be instructed by a workcard, if these are used, or orally if this is more usual. Every attempt should be made to keep this activity seeming the same as the others so that pupils don't suspect that it is a 'test'.
The teacher should have a list of things to look for in how this activity is carried out. For example:
- does the pupil choose bands of the same length when comparing different thicknesses?
- does he try all the different thicknesses there are?
- how does he compare the sounds?
- does he find a pattern or relationship?
- can he use it to make predictions?

In some cases it may not be safe to assume that what the pupil does is the result of deliberate thought. He may, for example, choose bands of equal length by chance or because 'they seem to fit the box best', rather than to control a variable. So it would be best for the teacher to talk through the investigation with the pupil, asking for reasons for his actions and justifications for his conclusions. If a relationship has been found then the teacher should have ready another rubber band of a different thickness and ask the pupil to *predict* the pitch of sound it will make when plucked.

Through these observations it would be possible to have some evidence of the pupils' awareness of variables, ability to make fair comparisons, to record results and find patterns in the results. Firm conclusions could not be drawn from one activity and it would be necessary to repeat the observations in other activities.

The process of assessing by observation

The above example reinforces the notion that observation doesn't just happen by being in the presence of children but may require certain actions of the teacher both in preparation beforehand and whilst in contact with the children.

The preparation beforehand is in terms of knowing what scientific development implies and being aware of the aspects of children's activities which will give evidence of this. *There is no avoiding the obligation to know what learning in science involves, for neither teaching nor assessment can be effective without it.*

Some of the more important positive actions involved on the part of the teacher are summarized in **Figure 9**.

Selection of criteria

For assessment of children's development in the various attitudes, skills and ideas, the information obtained by observing them has to be set against criteria, as suggested on page 33. Criteria can be expressed in a variety of ways. For example the following questions might be used in relation to the pupils' ability to observe:

- Does he make observations when his attention is drawn to specific points?
- Does he use several senses?
- Does he select observations which are relevant to the problem in hand?
- Does he note events or features in an appropriate amount of detail?
- Does he make records to aid his observations?
- Does he make quantitative observations where appropriate?
- Does he attempt to pick out patterns in the observations?
- *and so on*

These questions, arranged more or less in order of developing abilities in observation, could be used to scan information about a pupil's behaviour and indicate the point reached in development. Put in another way, it would be diagnosing what the pupil could do and what he could not do. But at the same time such questions direct attention to certain aspects of behaviour, those which enable these particular questions to be addressed. Inevitably, the narrower the base of the questions, the narrower the range of information gathered. For example, it is possible to think of an alternative set of questions about observation, such as:

- Does he produce neat and accurate drawings?
- Does he distinguish carefully between 'method' and 'observations' when writing about an investigation?
- Does he copy accurately from the board?
- Does he remember what he has seen demonstrated?
- *and so on*

These questions would focus attention upon quite a different set of activities. The contrast between the two draws attention once more to the value judgements which are an inevitable part of any assessment. Two different views of what is meant by observation are represented here; neither is 'correct' and both depend upon judgements as to what is educationally worthwhile.

Figure 9

Action	Comment
Providing activities in which pupils can apply or display certain ideas, skills, attitudes.	Obvious, but often ignored. Important to make sure that pupils have chance to use intellectual skills such as making predictions, creating hypotheses, before these are assessed, otherwise it may be wrongly inferred that they do not have certain abilities because there is no opportunity to display them. Careful interpretation of all information is required (see Chapter Six).
Discussing their activites with the pupils (dialogue).	Discussion in which pupils feel free to express their feelings, queries and ideas without the pressure of having to give the 'right answer' or feeling foolish if they don't understand something. Enables the teacher to see the pupils' way of looking at things and the interest they have in exploring various topics.
Asking questions which enable pupils to give their points of view.	The most helpful kinds of questions are 'open' ones, as opposed to 'closed' questions. These kinds of questions are well discussed by Tough in *Listening to Children Talking*, 1977, in relation to appraising language, but the same points apply to other intellectual skills. Another useful distinction in relation to questions is drawn by the Ford Teaching Project in *Self-Monitoring Questioning Strategies*, 1975, between 'subject-centred' and 'person-centred' questions. The former type focuses on the subject matter, while the latter asks about the pupil's ideas about the subject (e.g. *What do you think could affect how quickly the water dries up in a puddle* rather than *What affects how quickly the water dries up?*)
Listening to what pupils have to say.	An important part of both discussion and questioning. Teachers often do not listen carefully enough to what children are saying and miss its significance in telling them about the children's views and ideas. Information which could be used to help children often goes unheeded because the teacher is not really listening.
Noticing *how* they do their work.	Gives information about skills and attitudes which could never be apparent from looking only at the product, e.g. If pupils were using a hand-lens to look at the development of frogspawn before hatching, the resulting product (drawing or description) might lack detail because of the way the lens was used or because of misunderstanding of what to look for. Again, children do not always manage to put into practice what they intend to do and their planning may not therefore show in their product. Most important, attitudes such as responsibility and co-operation show little in the product but mostly in the way an activity is carried out.
Looking at the products of work diagnostically.	Still important, in addition to looking at the process, is looking at what children produce in the form of drawings, models, oral accounts, etc., especially to find indications of how they interpret their activities, how skills have been applied, etc., rather than to see if the result is correct or not.

Examples of criteria

One of the most interesting and early attempts to provide teachers with criteria to use in assessing their pupils was in the evaluation of the African Primary Science Project. The project was successful in stimulating active investigation in the children, but the methods of teaching were so new to the teachers that they needed some guidelines to help them find out if their pupils were developing along the intended lines. In response to this need the following were suggested as 'questions a teacher can ask himself as he watches a child's work from day to day':

1. Does he make suggestions about things to do and how to do them?
2. Can he show somebody else what he has done so they can understand him?
3. Does he puzzle over a problem and keep trying to find an answer, even when it is difficult?
4. Does he have his own ideas about what to do, so he does not keep asking you for help?
5. Does he give his opinion when he does not agree with something that has been said?
6. Is he willing to change his mind about something, in view of new evidence?
7. Does he compare what he found with what other children have found?
8. Does he make things?
9. Does he have ideas about what to do with new material you present to him?
10. Does he write down or draw some of the things he does, so he does not forget what happened?

. . . (Duckworth, *Evaluation of the African Primary Science Project*, 1970)

Criteria at a more detailed level were proposed by the Progress in Learning Science Project relating to each of the goals which were listed on page 24. In this case the criteria, developed with the help of several working groups of teachers (Elliott and Harlen, *Portrait of a Project*, 1980), were expressed at three levels of development for each of the goals. As examples, the criteria relating to the goals whose meaning was given on page 24 were as in **Figure 10** (the direction of development is from left to right).

The value of the criteria is that not only do they provide means of interpreting what children do in terms of their development, but they also suggest the kinds of observations to look for in the first place. They focus observation on to significant aspects of children's actions, products or speech. They provide the guidance which helps some teachers to see signs of difficulty or of progress in their pupils; signs which were there all the time but not previously recognized and interpreted as useful information.

Figure 10

OBSERVING

Makes limited use of his senses, noticing only some of the things which can be observed in the situation or only those which are pointed out.	Makes all kinds of observations, using several senses, though not able to discriminate the more important from the less important observations for the enquiry in hand.	Makes wide-ranging observations and can select from them the information relevant to a particular problem or enquiry.

CONCEPT OF CHANGE

Though he may notice changes which take place in living and non-living material he does not link them with associated circumstances or conditions.	He associates changes with the presence of certain features or conditions. Seems satisfied that the conditions account for the change and does not consider the interactions which take place.	Notices patterns in various kinds of change so that in most cases he distinguishes between broad categories such as growth, development, physical and chemical change.

OPEN-MINDEDNESS

Tends to stick to preconceived ideas ignoring contrary evidence.	Will take notice of some opinions and ideas different from his own but not others, being influenced by the authority behind alternative views rather than the strength of the evidence or argument.	Generally listens to and considers all points of view and relevant evidence, accepts new ideas if the evidence is convincing.

Information of this kind may be used almost immediately it is gathered, in the context of teaching. Often there is no need to record it in detail, but at the same time it may gain value by occasional systematic recording. The question of recording will not be pursued at this point, however, but is taken up in Chapter Six.

5 Summing up achievement

Introduction

It is not always easy to make a sharp distinction between the assessment considered in the last two chapters and that considered here, but broadly speaking the differences in purpose and properties of the information are as shown in **Figure 11**.

Figure 11

	Day to day assessment	*Summary assessment*
Purposes	To give information to guide the selection and progress of day to day activities.	To find out what pupils have achieved at a certain point in their course or in their develoment.
	To be used primarily by the teacher(s) of the pupil.	To inform other teachers, parents and perhaps outsiders to the school of pupils' achievement.
	To keep track of the progress of individual pupils.	To enable comparisons to be made (if desired) between pupils, classes, schools.
Properties	Must cover all goals.	Must include all main groups of goals but to cover all individual goals would mean too much testing.
	Must be sufficiently detailed to have a diagnostic function.	Groups of skills or abilities must be assessed thoroughly enough to give a reliable measure.
	Overall scores not relevant; concern with progress in various skills, abilities, attitudes.	Grades or scores required for making comparisons between individuals.
	Not necessary for all pupils to be tested in the same way.	For fair comparisons between classes or individuals, or to give standard information about individuals, all pupils have to be tested in the same way.
	Information best related to criteria, indicating what pupils can and cannot do.	

There are, of course, many similarities between these two types of assessment, the chief ones being that both should be designed to give as much information about as wide a range of goals as possible with minimum interruption to teaching and learning.

Purposes of summary assessment

There are two main types of purpose which this information serves: those relating to individual pupils and those relating to groups or classes of pupils. Summary assessment of pupils can serve to confirm or supplement the more informal judgements that teachers may make of their pupils' achievement over a certain period of time. This is particularly useful if pupils are to be divided into 'ability' groups or compared one with another for some other reason. This need not imply a competitive regime, but one which is attempting to 'match' pupils and classes to available courses and materials. Although the day to day assessment could be summed up and used for this purpose, there is often a preference among pupils, teachers and parents for some assessment which treats all pupils more obviously in the same way as each other than appears to be the case for informal day to day assessment. In fact the appearance of 'fairness' in a test situation may not be borne out when one considers whether in a standard set of questions pupils are given equal opportunity to show what they can do. Some combination of information of each type may well be better than either on its own.

Information about the performance of groups or classes of pupils may be used for a variety of purposes. The head of a department or school may wish to know the general level of achievement in different classes and follow this as a routine year by year. The decision may be to use tests made in the school, in which case any comparisons can only be made within the school, or to use tests which have been nationally standardized so that comparisons can be made between the performance of a group and a national, or a local, sample of similar pupils.

The growing pressure for accountability of schools is responsible for some considerable increase in testing for this purpose. This should be viewed cautiously. We must be careful that we do not over-test pupils or mistakenly assume that test information is the only kind which shows that a school is doing a good job.

A further purpose of summary assessment is to monitor the achievement of groups of pupils across the country as a whole, or a region or locality within it. This information may be used for policy decisions at a local or national level or so that the public at large can have information about what children in school can and cannot do. Normally the public does not have access to information about pupils' achievement and until recently there was no way of providing this information on a national scale. Since the Assessment of Performance Unit (APU) has been set up, however, the achievement in certain areas of the curriculum of national samples of pupils is being regularly surveyed and publicly reported. These results have little relevance to the immediate progress of individuals and indeed, as will be mentioned later, the surveys are conducted so that the results of individual pupils and schools are not disclosed. The aim is to find and report the range in performance of pupils in a national sample and in certain sub-groups (formed by region of the country, size and type of school, school locality, etc.). To meet this aim sufficiently large numbers of pupils have to be put in the same test situation and their results averaged; formal testing is the only feasible way of doing this.

The information required from summary assessment

What is required to serve the purpose of summing up achievement over a period of time is information covering the full range of goals of science education, but in less detail than for the purpose considered in the last chapter. For the present discussion, then, the information can be grouped under a few main headings, such as:
- skills of enquiry (observation, investigation, using equipment, etc.)
- skills of finding, interpreting and using information
- understanding and applying science concepts
- scientific attitudes

It might be argued that 'knowledge of specific facts' should be added to this list. This is very easy to test and there are plenty of examples to be found:

> *What is the boiling point of water?*
> **A**   0°C
> **B**   32°C
> **C**   90°C
> **D**   100°C
> (Kellington, *Assessment Questions for Integrated Science*, Book 1, 1979)

or

> *If both your parents have blue eyes, what will be the colour of your eyes?*
> *What is the name of the part in the centre of the eye, through which light passes?*
> (Progress Test, Level 2a 'Ourselves', West Sussex Science 5 – 14 scheme, *Science Horizons*, 1981)

The question which should be asked about items such as these is: do they really tell us anything important about children's scientific development? The knowledge of facts is a means to building broader generalizations, or concepts, which constitute the more important knowledge that children need to help them make sense of the world around them. Testing knowledge of specific facts must not be a replacement for testing the grasp of concepts and, more importantly, the ability to apply them in new situations. Moreover testing facts is dangerously easy and often replaces assessment of more central, but less readily tested, areas of achievement in science.

Methods

The framework for describing assessment methods, presented in Chapter Two, will again be used to consider the appropriate ways of summing up achievement. The discussion must be related to the purposes and desirable properties of the information which were mentioned on page 40.

Presentation
Presenting problems on paper, perhaps using pictures or symbols, is certainly the most convenient way of giving several pupils the same task in roughly comparable conditions. It is more suitable for giving information about some kinds of goals than others, however, so we should consider each separately.

Enquiry skills are concerned with the planning and carrying out of experiments or investigations. To assess this process *as a whole* it is necessary for pupils to be given problems to tackle in practice or a real experiment or investigation to perform. This requires a practical situation; the process of investigation cannot be adequately assessed on paper. Thus tests which are wholly written or which do not involve the pupils in handling and manipulating objects and equipment cannot assess the process of investigation as a whole.

However, the process of practical enquiry can be divided into component skills, such as identifying variables, planning procedures to control variables, taking measurements or making comparisons, making observations, recording findings, interpreting findings and drawing conclusions. Some of these can be assessed on paper quite adequately whilst others still require practical situations. The difficulty of setting up practical situations means that there is always a preference for written tests and inevitably these are sometimes proposed where they may well be inappropriate. In the following examples the question uppermost in mind must be whether what is tested is the *skill* or *knowledge about the skill*.

21 Look at this drawing of water in a measuring jar.

How much water is in the jar?

A 3 ml
B 8 ml
C 9 ml
D 13 ml

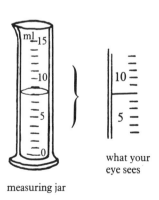

measuring jar

what your eye sees

23 Look at this drawing of a balance.

3 metal cubes each of 2.5 g are put on the balance.

What is the new weight on the balance?

A 2.5 g
B 18.0 g
C 22.5 g
D 25.0 g

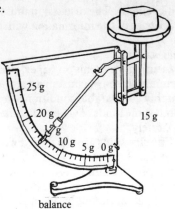

balance

Here is a ruler marked in centimetres (cm). Work out the length in cm for each black rod.

28 Length of rod is
A 4.0 cm
B 4.5 cm
C 5.5 cm
D 6.0 cm

(Kellington, *Assessment Questions for Integrated Science*, Book 1, 1979)

It could be argued here that knowing how to read a measuring cylinder and answering the written question successfully is not the same as reading a measuring cylinder in practice. It is certainly not the same as *using* a measuring cylinder, which involves knowing how to line up the eye with the meniscus and coping with the 3-dimensional complexity – aspects which are removed in the written form of the question.

Much better, but more troublesome in terms of preparation, for testing these abilities are the practical tests which are suggested by Kellington in the same book. These include preparing water in a thermos flask and asking pupils to measure the water temperature using a thermometer; collecting wooden rods whose length is to be measured; putting coloured water in a measuring jar or cylinder and directing pupils to measure its volume.

Written questions can be produced which test whether pupils can identify variables, control variables, plan a test, know what observations to make and how to use the results. The following examples show some of these.

In an experiment to investigate whether water is taken up by the roots of a plant, the bottle of water A was set up with the plant supported in it. A similar bottle B, containing water to the same level as in A was put beside it.

43. What was the reason for bottle B?

   ○ to record the level of water in A at the start

   ○ to see if the plant made any difference

   ○ to fill up A when the plant took in water

   ○ to keep the temperature of the experiment steady

   ○ none of these

(Brimer et al., *Bristol Achievement Tests*, 1969, Level 5, Study Skills, Form B)

11 Here is **'Sudso'** washing powder being tested against **'Dazzle'** to see which washes better.

front loader

80°C

Methods   45

Which one of these would you use (with **'Dazzle'**) to make the test a fair one?

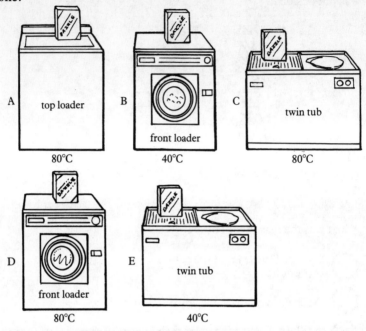

(Kellington, *Assessment Questions for Integrated Science*, Book 1, 1979)

Such questions are useful, since they do assess knowledge about variables without which pupils are unlikely to plan and carry out investigations effectively in practice. It seems important not to confuse this performance on paper with performance in real situations, however. In real investigations there is interplay between thought and action; sometimes the action may make it very obvious that a certain variable has to be controlled for a fair test, but in other cases the need for control may be confused by the practical details, and the influence of other facts may have to be decided on the spot. In a clean, clear-cut question on paper there are none of the problems which can be the greatest challenge to scientific thinking in practice. For example in the washing machine question above, there is no discussion of what could be washed to carry out the test, yet in practice this would make all the difference to a fair test.

So despite the undoubted usefulness of some existing tests, many aspects of enquiry skills are inadequately assessed by existing test materials and this must be mainly due to the preponderance of written tests. The case for conducting practical tests is strengthened by this examination of what is offered through written ones.

Turning to skills of finding, interpreting and using information it is appropriate for these communication skills to be assessed on paper since the information concerned is generally in written or symbolic form. Many examples can be given of questions asking pupils to record information. The first comes from a series of graded tests of general achievement. The question is read out to the pupils, who answer by ticking in their answer books.

Look at the graphs in the next box. The school nurse weighed Jane, Fred and Karen. Jane weighed more than Fred or Karen. Mark under the graph that shows that Jane weighed the most.

The pupil's page shows this:

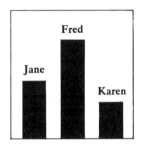
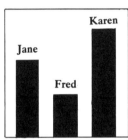
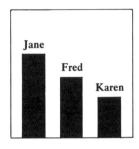

(Prescott et al., *Metropolitan Achievement Tests*, 1978)

An example of a more difficult question of this kind comes from the *Bristol Achievement Tests* for Level 5 (twelve to fourteen year olds):

These two diagrams marked A and B show how much of their wages a factory worker in India and a factory worker in England spend on certain things.

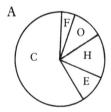
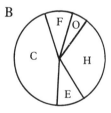

F – Fuel   H – Housing   O – Other things
C – Food and Clothing   E – Entertainment

61 The Indian worker spends more of his wage on

.... than he spends on all the rest.

Methods   47

62 Lower average temperatures in England than in India are mainly responsible for the English worker spending a greater proportion of his

wages on ....

63 The English worker spends as much on

.... as he does on ....

64 The English worker and the Indian worker spend the same proportion of their wages on

....

65 The English worker spends a lower proportion of his wage than the Indian worker on

.... and ....

(Brimer et al., *Bristol Achievement Tests*, 1969)

It is possible to design 'families' of questions around one initial problem or situation and this often helps to reduce the amount of reading or listening the pupil has to do to answer each separate question. Further, these families of questions can be devised so as to test different but related skills. For instance communication of information includes:
- reading information from various forms
- putting information into various forms
- interpreting information presented in various forms

One of the various forms might be a table, using keyed symbols, as is used in keeping records of weather observations. A table such as that shown in the next example could be presented, with the key.

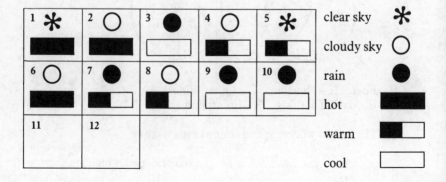

The following questions ask for different types of skill.
a (Reading from the table) Use the chart to describe the weather on Day 4.
b (Putting information into the table) Fill in Days 11 and 12 from these descriptions:
  1 Day 11 – a clear sky and warm
  2 Day 12 – cool and sunny
c (Interpreting information) Use the chart to decide whether these statements are true or not true:

	True	Not true
1 Whenever it rained it was always cool	☐	☐
2 Days with clear skies were all warm	☐	☐
3 All the cool days had rain	☐	☐

It was mentioned earlier that assessing grasp and application of science concepts is different from assessing knowledge of specific facts. The following are examples of questions which test applications rather than just the recall of the principles or facts involved:

8 In which drawing will the clothes dry fastest?

(Kellington, *Assessment Questions for Integrated Science*, Book 1, 1979)

16

Instructions to be read out: *You see three girls lifting a heavy box to the same height. Mark the space under the one who will be able to do it the easiest.*

(Prescott et al., *Metropolitan Achievement Tests*, 1978)

Why is it that dried milk does not go bad, but milk in a bottle will do if it is kept too long?
Because the things which make food go bad:
- ○ **a** need water to live
- ○ **b** do not like dried milk
- ○ **c** are found in water
- ○ **d** are liquid and not powdery

(Brimer et al., *Bristol Achievement Tests*, 1969)

When a force acts on an object the effect it has can be one or more of the following:

A. The object is made to start moving or increase its speed.

B. The object is made to slow down or stop moving.

C. The object remains at rest or moves steadily because the force balances other forces acting on it.

D. The size or shape (or both) of the object is changed.

32. Which one of these effects is involved when a bullet is shot from a gun?

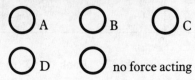

50   *Methods*

33. Which one is involved when a book is lying on the sloping surface of a desk?

○ A   ○ B   ○ C

○ D   ○ no force acting

(Ibid)

The fourth group of goals, scientific attitudes, are, by their nature, almost impossible to assess in test situations. Attitudes determine willingness to do certain things or behave in certain ways and, by definition, exist in the general pattern of behaviour in a variety of situations which call for a particular type of response. Thus it is not feasible to assess an attitude from responses to a single question or situation; all this would indicate would be that *in that situation* there was evidence of a certain attitude, but this would be no guide to general behaviour unless there was supporting evidence from responses in several other situations. So evidence about attitudes has to be picked up from a variety of situations. This means that it is very difficult to include attitude assessment in short tests, especially if these are written tests.

Unless a long test is constructed of questions specifically designed to detect attitudes – and this would be a dubious validity for young children – the assessment of attitudes must depend upon observing pupils' reactions during their activities. At present this area of assessment is undervalued and often dismissed because such methods are considered too subjective. However if some of the points made in Chapter Four, concerning the definition of criteria and the meaning of observation, are heeded, there is no reason why attitudes should not be included in summary as well as day to day assessment.

Ways of responding
In all assessment the influence on performance of abilities which are not under test should be minimized. So, particularly for young children, it is important that the demand of a question in terms of reading and writing should not prevent them showing what they can do in solving problems or showing scientific skills. Where possible the response should be made in the form of drawing, constructing or carrying out a measurement or investigation.

In written questions, one of the popular ways of reducing the burden of writing answers is to provide alternative answers from which children make a selection. Questions in this multiple choice form feature quite commonly in published tests because of the ease and uniformity with which they can be marked. However they have many disadvantages as well as advantages which should be considered.

A multiple choice question takes the form of a 'stem' which is a statement or question and several alternative answers from which the pupil has to select the correct one. For example:

Tigers are striped so that they can
A keep clean
B run faster
C hide easily from other animals
D be seen easily by other animals
(Kellington, *Assessment Questions for Integrated Science*, Book 1, 1979)

Some of the points in favour of such questions are:
- they do not require pupils to write answers and so do not limit pupils' performance on account of writing ability
- they can be marked easily; there is a key to the correct answer which can be used by unskilled markers or even by machine
- they can be answered quickly so that several such questions can be asked and a wide range of different topics covered

Points against questions in this form are:
- they can often be answered by intelligent guessing or elimination of ridiculous alternatives (distractors)
- the correct answer often sticks out because of significant words it contains or because of its length (in the tests from which the above example comes the alternatives are always presented in order of length to reduce the possibility of recognizing the correct one that way)
- they do not assess what pupils can *produce* as answers but only their *recognition* of right answers (plus other skills suggested in the two points above)
- they often require more reading than open-ended questions, which is time-consuming and a problem for some pupils (some multiple choice questions are given as group tests, presented orally by the teacher, as in the example above on page 47, thus avoiding this problem)
- they give no guarantee that the ideas behind the question have been grasped; the correct answer could be selected for the wrong reason
- they are very difficult to produce and should always be tried out on children who are not being assessed, to eliminate ambiguous wording or supposedly wrong answers which turn out to be right, or vice versa
- younger children find this form of question difficult to handle; they often tick more than one answer, however careful and explicit the instructions, which makes marking problematic (is the right answer plus a wrong one 'wrong'?)

One of the main reasons for the introduction of multiple choice questions was to avoid difficulties of marking. However, all that this did was to shift the balance of difficulty to the writing of the question. In the case of open-ended questions it is much easier to write a simple, straightforward question e.g. *What advantage do tigers get from being striped?* but more difficult to mark the answer. This straightforward question, however, does assess to a greater degree whether a pupil has grasped the concept to which it is directed. Ways in which such responses can be marked easily and reliably are considered under 'Representing results'.

Basis for judgement
As the purpose in this case is to give information about where pupils have reached in development in such a way that the progress of individuals or groups can be compared, then pupil-referenced marks or grades will not be adequate. It is generally helpful to relate the performance of all pupils to some criteria of performance so that the mark or grade has some meaning in terms of what they can and cannot do. This would be criterion-referenced. Referencing to a norm or average standard of the age group is also possible and useful for some purposes, particularly where there is to be a comparison of the performance of pupils in a school with the general level of performance of all pupils of the same age. However, the value of such comparisons is very limited, and for most purposes the result is more informative if it can be interpreted in terms of the development in relation to the goals listed on page 42.

Representing results
For written responses all types of marking (see page 17) are possible. It is usually only in the case of questions relating to skills of finding and using information that questions can be marked unequivocally right or wrong. For these questions there is often only a limited number of possible answers anyway and multiple choice or short answer format is useful. The examples on pages 47–48 show a suitable use of closed questions to which the answers can be marked right or wrong.

In the case of enquiry skills and application of concepts, however, it seems important to find out what children can produce as an answer rather than whether they can recognize a correct one or eliminate incorrect ones. Thus open questions are necessary, but present a marking problem. Inevitably answers given by children will be neither wholly correct nor incorrect. The most helpful way of dealing with marking in these cases is to use qualitative categories. For example, for the open form of the question about the tiger the categories could be:

a Answer which is regarded as correct and complete e.g. (according to age of pupils) it may be expected that it should contain references to:
- stripes mean tigers not easily seen
- therefore they can hide and hunt and not be attacked
- camouflage

b Answer which is correct as far as it goes but not complete e.g. only one of the above points.

c Incorrect answer, such as reference to stripes frightening other animals.

d Meaningless, or simply repeating the question.

The exact boundaries of the categories would be drawn according to the age of the pupils and the degree of sophistication in the answer which could be expected of them. So for some, a complete answer might be much less demanding than suggested here and for an incomplete one any reference to an advantage which in any way is correct would be enough.

When the general compass of the categories has been decided it is worthwhile filling them out with examples of answers which fall into one or another category. So, for instance, a decision would have to be made as to where an answer such as *It helps them to catch their food* would fall. By accumulating illustrative responses and having general descriptions of the categories it is possible to develop marking schemes which can be used reliably (i.e. so that an answer would be assigned to the same category by two different markers or by one marker on different occasions.)

The benefit of putting in this extra work is that the categories have much more value for teachers as information about what ideas pupils have developed. For purposes of forming an overall score from several questions, marks could be assigned to the categories (e.g. $a = 2, b = 1, c$ and $d = 0$). For other purposes there may be more interest in finding what proportion of the class give answers in each category, in which case numerical marks are not necessary. Furthermore a mark scheme of this same form could be devised for the application of other concepts, using similar criteria for what is regarded as complete and correct, incomplete and correct, incorrect, etc.

National surveys in science – the work of the Assessment of Performance Unit

As was mentioned at the beginning of this chapter, the Assessment of Performance Unit (APU) is carrying out regular surveys of pupils' performance in various areas of the curriculum and at various ages. Surveys began in 1978 when mathematics was tested for pupils of age eleven and fifteen. In 1979 language performance was tested in addition to mathematics at the same two ages. In 1980 the first science surveys were added and pupils at ages eleven, thirteen and fifteen in England, Wales and Northern Ireland were tested. Testing in mathematics, language and science was again carried out in 1981 and further tests are proposed as the programme of monitoring performance continues. In the short term future the performance of thirteen year olds in modern foreign languages will be assessed and, later, tests in technology may also be employed.

A recent pamphlet on the assessment of scientific performance at age eleven states that:

> The APU's function is to provide a general picture of the performance of school pupils across the country. The survey provides information about the whole age group by assessing a sample of it. It is not necessary to test

every pupil nor is it necessary that every pupil tested is given the same questions. The number of questions used in the monitoring can therefore be far greater than could be given to any one pupil and so can cover a wide range of performance. There are several different packages of questions, each of which is given to a sub-sample chosen so that its results are representative of the whole age group. Clearly, then, the results for any one pupil, or from any one school, cannot give a complete picture for that pupil or school. Results of individuals do not have any meaning until added together with others. To ensure that results cannot be misused, they are collected in such a way that the identities of pupils taking part are not known to anyone outside the schools and the results of individual schools involved cannot be identified.
(APU, *Assessing Scientific Development at Age 11*, 1981)

The framework set up to assess scientific performance consists of six categories which are divided into sub-categories (**Figure 12**).

Figure 12

Main categories	Sub-categories
1 Using symbolic representations	Reading information from graphs, tables and charts Expressing information as graphs, tables and charts Using scientific symbols and conventions
2 Using apparatus and measuring instruments	Using measuring instruments Estimating quantities Following instructions for practical work
3 Using observation	Using a branching key Observing similarities and differences Interpreting observations
4 Interpretation and application	Describing and using patterns in presented information Judging the applicability of a given generalization Distinguishing degrees of inference Making sense of information using science concepts Generating alternative hypotheses
5 Design of investigations	Identifying or proposing testable statements Assessing experimental procedures Devising and describing investigations
6 Performance of investigations	

(DES, 1981)

Separate sets of questions have been constructed for each sub-category and examples are to be found in the published reports of the surveys at age eleven, thirteen and fifteen (DES, *Science Performance at Age 11: Report No*

1, 1981, . . . *at Age 13*, 1982, . . . *at Age 15*, 1982). The methods of assessment vary according to the type of performance being assessed. Written tests are used for categories 1, 4 and 5 and practical tests for categories 2, 3 and 6.

The written tests are mostly open-ended, requiring children to produce either short or more extended answers. The following is a question assessing the construction of a graph.

Richard measured his bean plant every week so that he could see how fast it was growing.
He started (0 weeks) when it was just 5 cm high.
These were the heights for the first 4 weeks:

      0 weeks – 5 cm
      1 week – 15 cms
      2 weeks – 30 cm
      3 weeks – 40 cm
      4 weeks – 45 cm

Draw a graph to show how the height changed with time.

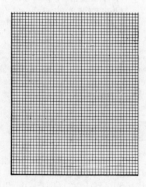

(DES, *Science Performance at Age 11: Report No 1*, 1981)

A few questions are multiple choice, such as the following one, concerning the ability to identify when assumptions are being made.

Look at this picture of people walking through fields in which there are animals.

Read the statements below.
Tick the one statement which you can be most sure is true just by looking at the picture.
- [ ] The walkers have left the gate open.
- [ ] Some sheep have strayed in with the cows.
- [ ] The farmer is going to be angry.
- [ ] There are some sheep in both fields.
- [ ] The cows will go into the other field.

(Ibid)

Some questions ask pupils to create hypotheses and, in marking, credit is given for any suggestions which fit the data given in the question and accord with accepted science concepts. The following is an example.

Walking along this footpath Thomas noticed that there was ivy growing on the trees but only round three-quarters of the trunks. None of the trees had ivy growing on the side nearest to the path.

Think of *two* different reasons why the ivy might grow only on some sides of the trees. Write the first at a) and the second at b)

a) I think it might be because ....
....
....
....

b) Or it might be because ....
....
....
....

(Ibid)

The practical tests are of two main types, those given to a group of pupils at a time and those given to individual pupils. For age eleven pupils the tests for 'Using Observation' are given to a group of eight pupils at a time. The pupils answer the questions individually in their own test booklet but the questions are presented orally to them all at the same time. The children are given objects to handle, photographs or drawings to look at, or are shown films, in the course of the presentation of the questions. For older pupils at ages thirteen and fifteen, the group tests take the form of a 'circus', with nine pupils circulating nine 'stations' at each of which a practical activity is presented. At all three ages the 'performance of investigations' is assessed through individual tests presented by a tester who observes the way the pupil carries out an investigation and discusses it with him. The performance is recorded on checklists.

All the practical tests are administered by teachers, temporarily absent from their schools in order to be trained and carry out the work. Thus the methods can be elaborate and intensive since the testers can give their whole attention to the work. In these conditions the assessment of practical work can be carried out thoroughly and the methods devised have made a significant contribution to assessment techniques. Teachers may not be able to use them in the same way as in the surveys – and it would not necessarily be wise to do so, given the very different purpose of the APU's assessment in schools – but there may well be things to be learned from the techniques and questions used which could be adapted for in-school use.

6 Keeping and using records of assessment

Are records really necessary?

The honest answer to this question is 'yes – but only if they are good records'. Some records, no doubt, are kept for the sake of keeping records and are never used. Clearly this is largely a waste of time, and is something which should be avoided by making sure that the form and content of a record is designed to serve the intended purposes. So first we should define a good record.

A good record:
- includes all the information relevant to serve its purpose, and is detailed when the purpose demands detail but no more than necessary
- makes a significant contribution to planning or decision-making at some level, or provides information which people want
- is not disproportionately time-consuming to prepare in relation to its value
- is rewarding to prepare in that the process of keeping the record is informative in its own right, in addition to having value in building up a useful product

Records which meet these criteria are a necessary part of teaching since:
- it isn't possible for teachers to keep in their heads all the information needed to help children or to plan suitable work for them
- it is not only the pupils' teacher who needs information about their progress, but other teachers, especially the head teacher, as well as the parents and the pupils themselves; some form of record is necessary to pass on this information
- bringing together information collected at different times and from different sources reveals patterns of change over time and patterns in behaviour which are very helpful for informing teaching decisions
- completing records provides the incentive for teachers to review what they know about their pupils and to realize what they do not know
- it follows that recording is an aid to assessment, since it prompts the gathering of information which is missing and facilitates reflection upon and interpretation of the information gathered

These points, of course, are equally true for all kinds of records in all parts of the curriculum. The value of records is well appreciated in mathematics and reading, and most primary and middle schools have quite extensive record systems for these. A recent survey (NFER, *Record Keeping in the Primary School*, 1980) showed, however, that in the case of science in primary schools the reverse is true, and that most schools did not keep records of pupils' achievement in this area. This did not mean that individual teachers may not have kept their own notes to help their planning, but these were not used to feed in information into the pupils' records. There are not, therefore, many good examples of record systems in science to follow and it is necessary to build up ideas from considering the pros and cons of the possibilities.

Types of record

Free comment
This can mean anything from a terse 'a good term's work' as a summary record to a detailed description of a pupil's responses to certain activities, often accumulated from shorter entries made regularly. In some infants' schools it is customary for each teacher to write records each day, noting anything which has struck them as remarkable about any child. Usually only five or six entries are made each day, but this builds up to a considerable bank of information over a time. There are children for whom there are many entries and some for whom there are few, but the process of keeping and reviewing the records draws the teacher's attention to both groups: 'Why have I noticed so much about Tim and Vincent?' and 'Why have I noticed so little about John and Gary?'

The strength of the completely open record is that it allows all kinds of comments to be entered, unbounded by headings or items to check against. The result can give a more complete picture of a child than a list of grades, ticks on a checklist or test scores.

This strength can, however, also be a weakness for certain purposes. Without any guidance as to the points to cover it will be inevitable that some aspects of behaviour or development are noted for some children and quite different ones for others. This makes it impossible to make comparisons between children or to combine the records of individuals into a class record if required.

So records in the form of free comment can be of value for teachers' private records, when they already know or can easily obtain the information not recorded, but they are best used as a supplement to other forms of record if they are to be used for purposes which require some uniformity between what is recorded for all pupils.

Lists of activities
These consist of lists of topics or activities, often relating to a series of workcards, which may be prepared by the teacher or are supplied with a programme of acitivities, as in the case of the *West Sussex Science 5-14* scheme. The latter simply lists the activities in the unit and instructs the teacher to 'Tick each activity when completed' for each child. This approach is sometimes questioned since it isn't always clear what 'completed' means. Some teachers, therefore, use a more elaborate system of recording activities

with different symbols to represent 'attempted', 'completed' and, sometimes, 'understood'.

The advantage of this type of record is that it is reasonably objective. It indicates the kind of things the pupil has encountered – he will have used certain equipment, been able to make some observations and possibly will have made some record of what he did. Clearly it is important to have this basic information recorded, if only to avoid repetition in the pupil's experience.

However there are disadvantages if this is the *only* form of record kept. It is inadequate on its own to convey whether the child just went through certain motions in 'completing' the activity, or whether he extended it, used his own ideas or learned something from it. It is more a record of the use of activities than it is of the learning of children, and we must remember that changes in children are the target of teaching and therefore of records.

Checklists of development

Although lists of activities could be described as checklists, the ones considered now are lists of skills, concepts, facts, attitudes, etc., which describe what might result from the children's activities.

An example is provided by the checklists produced by teachers helping the Progress in Learning Science Project, which produced the criteria for assessment by observation discussed on page 39. A method of recording progress was provided by linking the three statements to five boxes, as follows (an example not given earlier):

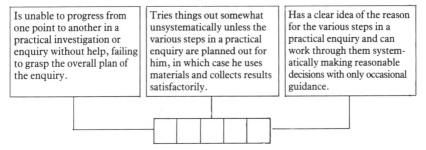

One way which many teachers have found useful is to make use of the five boxes under each set of statements in the checklist. The idea is that each box represents a point of development, either one indicated by one of the statements or a position in between two statements. Thus if the first statement describes a child's behaviour the record is made like this:

If the child seems somewhere between development described by the second and third statements it can be recorded like this:

▨▨▨▨▨☐   or like this   ☐☐☐▨

Subsequent records from the same child can be made in the same record sheet by blocking in changes in a different colour, e.g.

First record   ▨▨▨☐☐

change at a later time   ▨▨▨■☐

(*Match and Mismatch: Raising Questions*, 1977, p. 237)

In this way a profile for each child can be constructed, showing change over a period of time.

Profile of Simon

Autumn 1973 ▦
Summer 1974 ▨

Observing	▦▦▨☐
Raising questions	▦▦▨☐
Exploring	▦▨☐☐
Problem solving	▦▨☐☐
Finding patterns	▦▨☐☐
Communicating verbally	▦▨☐☐
Communicating non-verbally	▦▦☐☐
Applying learning	▦▨☐☐
Concept of causality	▦▨☐☐
Concept of time	▦▨☐☐
Concept of weight	▦▨▨☐
Concept of length	▦▨▨☐
Concept of area	▦▨☐☐
Concept of volume	▦▨☐☐
Classification	▦▦▨▨
Curiosity	☐☐▦☐
Originality	☐☐▦☐
Perseverance etc. . . .	☐▦▨☐

(*Match and Mismatch, Raising Questions*, 1977, p. 55)

The advantage of this record is that it condenses a great deal of information into a small space, making it easy to see patterns in the profile and in progress from one time to another. Each box is *not* a point in a five-point scale but represents a stage in the development towards one of the goals described by the criteria. It is important not to divorce the record from the criteria since the meaning would then be lost.

The record is not as time-consuming to make as at first appears. The important part of the assessment is the making of the observations, and if these have already been made then entering them in the checklist need take no more time than writing comments on each child. It need not be done more than two or three times a year, though some teachers have found it useful to use the boxes as a way of recording more regularly by putting a dot or cross in the appropriate box to record a significant event i.e.

Thus the record is gradually built up showing the scatter or range of behaviour in different activities as well.

The disadvantages of this type of record lie partly in the danger just mentioned, that it will be seen as a five point scale. The equal spatial intervals might easily be interpreted as equal intervals in development, or worse, in performance, whereas this is not the case. The relationship between the three criterion statements for each goal is one of supposed order only i.e. the third statement indicates further development than the second and the second further than the first. A further disadvantage is that the record is too detailed for some purposes. For summary purposes, it is not necessary to assess all goals by observation and some selection could be made; the concepts in particular, could be assessed in other ways. For this assessment, the profile could be condensed under fewer headings, perhaps the four suggested on page 42.

Ratings or grades
The most familiar form of this type of record is an A → E grading, either as a record of an overall assessment, or as a series of grades against headings which may be fairly general ('concepts', 'practical work' etc.) or a series of more detailed goals. The record sometimes takes a different form, as a rating from 1 → 5, or as a dimension, such as described in the example on page 21.

co-operative ☐☐☐☐ not co-operative

The number of grades or levels varies, the range usually being from 4 to 7. It seems strange that use of this method of recording is widespread despite its obvious shortcomings. Its attraction must lie in its apparent economy and precision. However, a closer look suggests that it is false economy and spurious precision. Even though it does not take much time or effort to make the record, even that much is wasted if the result can be misinterpreted because there is no clear indication of the meaning of an A, B or C, or a rating of 3, 4 or 5. The absence of criteria is sometimes explained on the grounds of an implicit norm-referencing i.e. that C (or 3) means 'average', and that points either side show a smaller or larger departure from average. This does not however clear up the uncertainty, but rather shifts it to the identification of the 'average'. For some purposes, as we have seen, it is more important to know what a child can do, and what ideas he has already developed, than to know how he relates to some kind of average. Thus, unless there is some definition of each grade or rating point which says what it means in terms of pupils' abilities, the precision of the record is more apparent than real.

Test scores

Scores in tests derive meaning in terms of achievement from the content of the test. So if the record of a school-produced test is 6 out of 10 or 25 out of 30, the way to find out what this means as far as what the pupil can do is to look at the test. It should be routine for copies of written tests to be kept with results so that the meaning of the marks is not lost. The same applies to tests published with curriculum materials which are not standardized tests.

Tests which have been standardized have been given to a large and, hopefully, representative sample of the age group and so the 'average score' or norm has a more precise meaning than is the case where a teacher can only judge the average from her own class or school. The particular advantage these tests have is for purposes which require comparison of performance between individuals or groups in one school with the national or local average. The need for such comparisons does not occur very frequently, however, and for purposes relating to teaching, it is more useful to look at the test as indicating what the pupil can and cannot do, than as just a means of producing a percentile score.

Records for different purposes

In devising forms of record it has to be acknowledged that the record can only reflect the information which has been gathered. A good record system cannot make up for a poor assessment system. On the other hand, it may be true that the good assessment system can be limited by the records which are kept. For example, a great deal of varied information can be gathered by observation but if this is to be condensed into a naïve rating instead of being linked to criteria when recorded, the richness cannot be recovered from the record.

Records form a communication link between one point in time and another, and between one person and another. What is important is that the meaning of the words, marks, symbols used by the creator of the record should, as far as possible, be able to be recaptured by the reader of the record.

Records also have a permanency which is often a disadvantage unless they are kept, and updated, regularly. Information which is out of date may be more dangerous than no information. These points have been borne in mind in suggesting the appropriateness and feasibility of types of record for different purposes, as follows.

For day to day assessment
- Record of development kept against detailed checklist of goals either as comments or as ticks related to descriptive criteria
+
- Record of activities attempted with comment or success/extension/interest/ideas contributed, etc.

For summing up achievement
- Record of development in relation to main areas, e.g.
    enquiry skills
    communicating information
    science concepts
    scientific attitudes
  either as comments or as ticks related to descriptive criteria.
+
- Record of broad topic areas covered (name of curriculum materials used, if appropriate)
  + (optional)
- Comments on aspects not covered by above
  + (optional)
- Test scores with tests attached or named (if standardized tests)

Reviewing and improving records

The form and content of records must be reviewed and revised to keep in step with changes in approaches and ideas in primary science teaching. It seems entirely appropriate that there will always be development of new ideas about content and methods if science teaching is to serve its central purpose of helping children to understand the world around them. For young children this world is fairly small, beginning with their immediate surroundings and gradually expanding. But even this intimate world is changed by, for example, the growing recognition of the importance of technology and the increasing prevalence of the microcomputer. These, and other changes in emphasis or concern in everyday life, quite rightly should be reflected in the developing curriculum, and therefore in the content of the records which are kept.

It is not only the content but the form of records which should be kept under review. It is not easy to arrive at a record system which suits all those who prepare the records and all those who use them. Teachers often try different ways of keeping their own records for day to day planning in the hope of finding one that is more efficient. It is useful to *evaluate* records regularly since this can suggest modifications for improving existing methods.

The process of evaluation means gathering information about something, and defining criteria to decide how satisfactorily or otherwise it appears when judged against these criteria. More will be said of this matter in the next chapter. For the moment it is enough to state that for evaluation of records we must obtain *information* about them and then apply certain *criteria*. The extent to which the information meets the criteria then suggests the changes which might help to make the records match up better to the 'ideal' reflected in the criteria.

Information

A simple list of what might be included can cover records kept for both day to day and summary assessment, though clearly some information is less relevant to one than the other. Information should be gathered to answer the following questions:

a What features (skills, concepts, attitudes) are covered by the record?
b How detailed is the information about each feature?
c Does the record contain within it any indication of recent progress and the direction of future development?
d How often in the year, and when, are the records consulted?
e Who consults the records?
f Who completes the records?
g How often, and when, are entries made?
h If more than one teacher is involved, do they consult each other and agree an entry?
i Are there guidelines for completing the record including definitions of terms and meanings of grades or ratings (if used)?
j Are tests included with test scores?

Criteria

The information about the record should be scrutinized to see how well it matches the criteria for the 'ideal'. It must be stressed that there are no absolute ideals and the following suggestions obviously reflect a certain value position about the role that records should play in education. They are intended only as examples of an approach to evaluation.

It is best to consider separately the records for different purposes. Firstly day to day assessment records, for which questions **e**, **f**, **h** and **j** are not relevant. The criteria against which the information is judged might reflect an ideal, in that:

a they include some way of keeping track of all major goals relating to enquiry skills, science concepts and scientific attitudes
b they enable a picture to be built up of each child's strengths and weaknesses
c they indicate where a child has not made progress and may need help, and the direction in which he may be ready to take further steps
d they are used regularly in planning activities
g some record is made of activities carried out or of skills shown by pupils at least every two weeks
h the terms used in the record are defined so that the meaning of each feature of development included is clear, and any ticks or symbols have consistent and clear meanings

For summary assessment we could ask how well the records match up to the following criteria, to see if:
a they include all major types of goal
b information required can be easily picked out
c cumulative records are summarized in a way which shows the progress which has been made
d records are consulted at the beginning of the year in preparation for reporting to parents and at any other times when they could throw light on a query about a pupil
e the records are consulted by all those who can use the information in them (e.g. head teacher, new class teacher, new science teacher)
f the record is completed by the teacher who has been most concerned in the science activities
g entries are made at the end of each term or each year as considered appropriate
h the record reflects the combined views of all those who have been involved in the science activities
i the terms used in the record are defined so that the meaning of each heading is clear (e.g. if one is 'skills' there should be a list somewhere showing which skills are included), and any grades or ratings are defined
j where test scores are recorded the tests from which they were derived are appended or named (if published tests).

Interpretation of records

The point of keeping records is that they inform decisions about future action. They may indicate that progress is fine and that the only action required is to proceed according to plan. More often, though, they will indicate a possible problem: some children will not be responding and making progress as expected; perhaps it seems at times that no-one in the class is ever going to suggest an investigation or propose an explanation which is testable. The job of teaching is to overcome such problems, and if they did not occur one might suspect that the pupils were insufficiently challenged by their activities.

Records can be used to suggest explanations of the information they convey. If one child is showing no evidence of progress in 'interpreting observations', for example, it is helpful to scan the records of other children to see if it is a general feature, or confined to the pupil in question or perhaps to his working group. If it is something which does not seem to be general then the child's records might be examined more closely. Has he had opportunity in the activities he has carried out to find patterns in observations? Has his ability in observation or communication shown lack of progress? The search for an explanation will naturally go beyond the records to include more detailed observation of the way he goes about his activities.

One child with such a problem was found to be depending very heavily on members of his working group, who unwittingly forced upon him their own interpretations of what they were doing, which he became used to accepting without understanding. This child benefited from rather simpler activities for a while to build up his ability and confidence in 'putting two and two together'. In another case a teacher realized that *she* was responsible for one group having little opportunity to interpret their findings because she tended to jump in with too much help, leaving the pupils little to do except agree with her.

Where there appears to be a pattern in the records across the whole class, or a group, it is clear that the problem is likely to be rooted in the opportunities which are provided for the pupils rather than in some inherent inability in the children. The same argument applies to the individual pupils as well; the first response should be to ask: Is he getting the opportunities to progress?

The focus of attention then turns from the assessment of pupils to the assessment of their learning opportunities in science. Some ideas for carrying this out are contained in Chapter Seven, where the ideas introduced for looking at records are extended to the evaluation of other factors affecting teaching and learning.

7 Evaluating opportunities for learning

Introduction

A discussion of assessment which dealt only with the assessment of pupils would be very definitely incomplete. As we saw in the last chapter, when the records of assessment of pupils are put to use in finding how to improve learning then it becomes obvious that it is necessary to know about what has been provided to encourage learning.

It could be said that the assessment of learning opportunities logically comes *before* the assessment of pupils, on the grounds that there is not much point in looking for products if there has been no process which could provide them. But the relationship between assessing what children learn from their activities and the opportunities for learning which the activities provide is not a simple one. Opportunities for learning are determined by the nature of the pupils as much as by the activities. That is, the same activities may provide rich opportunities for learning for some pupils but poor ones for other pupils. Thus it is necessary to know something about the pupils – to assess their skills and ideas – before the appropriateness of the activities can be judged. At the same time it is necessary to know about the activities – to assess their potential – before their impact on the pupils is assessed. This cyclic relationship is represented in **Figure 13**.

Figure 13

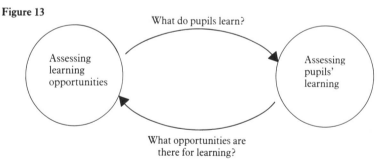

There is a further reason for assessing learning opportunities relating to those goals of teaching whose achievement is very difficult, and in some cases impossible, to assess in pupils in the short term. In all areas of the curriculum there are goals about affecting pupils' future values and attitudes and their behaviour in everyday life which, by definition, cannot be assessed in school. In science such goals would concern 'continued interest in the role of science in everyday life', 'responsibility in the use of scientific knowledge' and the Science 5–13 objective 'willingness to extend methods used in science activities to other fields of experience'. For these goals there is little that can be assessed in school science that relates to their achievement, yet most teachers would subscribe to their importance. What we *can* do, though, is to ensure that the pupils' experience includes opportunities to develop these attitudes. We cannot be sure that the opportunities will inevitably lead to the desired outcome, but we can be sure that if there are *no* opportunities there is no chance of the learning taking place.

The meaning of evaluation

The meaning of assessment was discussed at the beginning of Chapter Two, where it was defined as a process of gathering information. Here the wider term 'evaluation' is more appropriate since the concern is with information gathered by a variety of methods (of which assessment is only one) and about a variety of features of the teaching and learning activities. As mentioned in the last chapter, evaluation is a process in which information is judged against various criteria (standards or expectations). This description of evaluation and its relationship to assessment is shown diagrammatically in **Figure 14**.

Figure 14

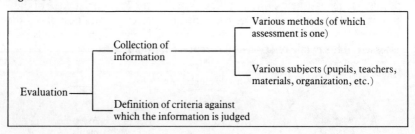

Imagine the information about the class organization given here to be judged against two very different sets of criteria, A and B, as in **Figure 15**.

Figure 15

Criteria A	*Criteria B*
A good classroom organization is one where: ● the desks are arranged in rows ● the desks face the blackboard ● the pupils remain at their desks ● the pupils all carry out the same activities at the same time ● the pupils cannot easily talk to one another	A good classroom organization is one where: ● the pupils can move about freely ● the desks are arranged to promote group working ● there can be many different activities going on at one time ● pupils can discuss their work with each other ● pupils help to decide where and how they carry out their work

70   *The meaning of evaluation*

This makes explicit a relationship which is often not spelled out but is present and effective none the less. Whenever an evaluation is given, formally or informally, the person giving it has in mind information about the subject and is using certain criteria to give a judgement about it. The same information could be judged quite differently if different criteria were used. Take, for example, the information about class organization which is contained in this scene (information does not have to be verbal).

There is little doubt that the judgement as to how 'good' is this classroom organization would be very different according to whether criteria A or criteria B were being used. The judgement itself is of little significance unless one knows what the information was and what criteria were used in making the judgement.

So, for evaluation, we have to be concerned with the collection of information and the spelling out of criteria. These two parts of the process will now each be considered.

Information about learning opportunities

If we are to consider opportunities for all kinds of learning – enquiry skills, science concepts and scientific attitudes – this means that the information must concern *what* activities are carried out, *how* the activities are carried out and the *organization* in which they take place. Thus the content, teaching methods and organization form the main headings under which information about 'what is done' should be collected. For each of these, a description is required, in whatever form is most appropriate, which is as

The meaning of evaluation   71

complete and unselective as possible. (Of course no information can be completely objective – any information is always a selection from the full range of possibilities. This cannot be avoided and all we can do is to guard against the bias it could bring; recognizing the problem is an essential part of this process).

To collect the information two main decisions have to be taken. Firstly the time period which is to be covered and secondly whether the activities in that time period are to be sampled or treated as a whole. The decision about time must take into account the need to have a long enough period for the activities to be typical and to include some attempt to cover the major goals, but at the same time a short enough period to be handled. A whole year is ideal from the point of view of an adequate sample of time, but is really too long for detailed study. A term is a good compromise, though a half-term would be adequate if it were felt to be representative of the methods, activities and organization generally used.

The time span affects the decision about sampling, since the shorter the period the more chance there is to cover all activities. When groups of pupils carry out different activities there is another aspect of sampling to consider, whether to follow just one group or include all the activities going on. This may be decided by the purpose of the evaluation; for example the interest may focus on the learning opportunities of one group or one pupil in order to throw light on lack of progress of a certain kind. It may focus on the whole range of activities going on in the class if the purpose is to find out if this has any deficiencies in relation to all the main types of goal. The decision must depend upon particular circumstances and purposes. In the following 'the activities' refers either to a sample or the whole set of activities experienced by either one group or several groups over the period of time considered for the evaluation.

For each of the activities the aim is to describe what happens under the three headings (**Figure 16**):

Figure 16

The activities	What was done		
	Content	Teaching methods	Organization
Activity 1			
Activity 2			
Activity 3			

This is not a data collection form but a framework, which will be extended later.

Content

Information about the subject matter encountered can be collected by a variety of means: from examining any workcards or other written materials used by the pupils, the pupils' written records or drawings, models or any other products, or from the teacher's notes (if any) made at the time concerning what the pupils actually did (not what was planned). More information can come from talking with the pupils, asking them what they did.

Teaching methods
It is less easy to collect information in retrospect about the methods used unless the teacher made notes at the time or was observed by someone else. Often it is more helpful to sample a short period of teaching and observe it in depth with the help of someone else, or a tape-recorder, than to attempt to generalize about many activities. In this way detailed information can be obtained about:
- the kinds of questions asked by the teacher
- whether the pupils or the teacher suggested how to interpret, record, extend an investigation
- what questions the pupils asked
- what answers the teacher gave
- how pupils' activities were decided and begun
- what kinds of help were given to pupils who were stuck
- *and so on*

A small portable tape-recorder can give a teacher a great deal of information about her role in the pupils' activities. Other methods can be used, however, if there are opportunities to have an observer to help or to make a video-recording of a lesson. If these aids are out of the question – as they usually are in the normal run of events – then it is still possible to obtain useful information from the children. Talking over an activity with a group of pupils can lead to indications of the kinds of help they received, whether they were using their own ideas or following directions, what kinds of help they would like to have had, whether there were alternatives open to them which they had to consider, whether an interpretation of findings or a conclusion was their own or given to them. All this has a great deal to say about the methods of teaching which were used.

Organization
This refers to the arrangement of the physical surroundings and to the organizational structure of the class. Part of this information might best be provided by photography or classroom plans. The arrangement of desks or tables, the position and organization of materials and resources, the areas for display, the location of reference books – all these have a significant part to play in determining the opportunities there are for learning of various kinds.

The other part of the information concerns the organization of pupils. This is generally decided by the teacher and information is immediately available to answer questions such as:

- do the children work in groups, individually, or as a whole class?
- if there are groups, how are they decided?
- do all pupils work on science activities at the same time?
- how do they obtain materials for their activities?
- where do they work (always in the classroom, in a resource area, science room or corner, elsewhere)?
- how long do they spend on science activities?
- *and so on*

Criteria for evaluation of learning opportunities

When information about what the pupils have done has been gathered and brought together, the next step is to examine it critically for where it falls short of intention or to what extent it provides the experiences thought to be desirable for the particular pupils involved. For this the events which occurred have to be compared with criteria reflecting intentions in relation to all the goals, reflecting what is considered appropriate for the pupils and representing a kind of desired 'ideal'. These further steps are included in an extended version (**Figure 17**) of the framework given earlier:

Figure 17

The activities	What was done			The criteria reflecting the 'ideal'	Judgement of adequacy and suggestions for improvement
	Content	Teaching methods	Organi- zation		

Remembering that the criteria will be highly dependent upon value judgements (cf the examples for class organization given on page 71), the examples given below *must* only be taken as illustrative of the method and not indicative of the criteria to be adopted.

There are several ways of expressing criteria and of applying them. They can be expressed as statements of the ideal, as in Chapter Six, or as questions which, as it were, interrogate the information:
- were children able to take responsibility for looking after living things?
- were their ideas for planning investigations met with encouragement and praise?
- were they able to work together in a group and help to co-operate with each other?

They might also be expressed as events which are considered to be 'critical indicators' of intended opportunities:
- pupils being allowed choice of what they do or when they do it

- pupils being able to collect and replace materials as they need them without waiting for the teacher
- books being available for children to broaden their knowledge of the topic being studied.

One of the most useful approaches is to link the criteria to the goals which it is hoped to achieve through the experience of science activities. All kinds of goals should be included: those for which there is some hope of assessing achievement and those which defy assessment by their nature. In this way there is a consistency in the processes of planning and evaluating and the evaluation should help to improve the planning and execution of the plans.

So it is appropriate, for the consistency of ideas underlying this book, to return to the goals which were listed in Chapter One. A sample only, of two enquiry skills, two attitudes and two concepts are considered. For these examples some suggestions are made as to criteria which could be developed and used to evaluate what has been done in the activities over a certain chosen period of time. In each case the criteria would have to be interpreted in relation to the development of the children.

Criteria relating to 'observation'
The ideal curriculum would provide opportunity for:
- watching certain events (e.g. birds feeding), discussing what is seen and realizing how much can be found out by systematic observation
- looking in depth at an object or event and noting (or describing orally) all that might be relevant to some enquiry surrounding it
- looking for similarities and differences between objects or between different points in a succession of events
- using the senses of touch, hearing and smell as well as sight in exploring events
- making as many observations as possible about an object and then deciding which are relevant to helping a particular enquiry concerning it
- grouping objects together and explaining the basis of the grouping
- using observation of given characteristics to classify objects
- learning how to use instruments to extend the power of the senses
- using such instruments when appropriate to aid observation (e.g. a hand lens to watch snails eating or to look at the shape of crystals)

Criteria relating to 'identifying and controlling variables'
The ideal curriculum would provide opportunity for:
- identifying all the things which might be influencing a particular event (such as how high a ball bounces)
- identifying all the ways in which an object could be changed without affecting its function

- discussing the 'fairness' of comparisons which are made between objects or events
- systematically investigating the effect of all the things which differ between two objects being compared
- exploring variables in situations where they are few in number and easy to control and to vary
- discussing the possible influence of variables which have not been controlled (i.e. being allowed to make mistakes but learning from them)
- using measurement where possible to compare the effect of variables

Criteria relating to 'perseverance'
The ideal curriculum would provide opportunity for:
- finishing each activity to each pupil's satisfaction
- obtaining the materials and resources necessary to complete a task to each pupil's satisfaction
- beginning an investigation again, using a better approach, without a sense of failure
- help in finding alternative ways to tackle problems
- reinforcement of perseverance by praise and example from the teacher

Criteria relating to 'co-operation'
The ideal curriculum would provide opportunity for:
- working together with others, sharing space, materials and equipment
- joint planning among members of a group on what is seen by them as a shared enterprise
- participation in decisions about how materials and resources are to be shared in the classroom
- discussion which helps pupils to appreciate and consider the needs of others as well as themselves
- reinforcement of co-operation by praise and example from the teacher

Criteria relating to general concepts
Whilst the four previous examples have concerned chiefly the methods of teaching and the organization, these most clearly relate to the content of activities. The criteria for each are fewer in number and might be considered in relation to groups of concepts rather than to each of the detailed statements on page 10.

About living things
The ideal curriculum would provide opportunity for:
- the complete life cycle of some animal to be seen at first hand
- the stages in development of some plant to be seen at first hand
- taking responsibility (under supervision) for the care of living things in the classroom

- visiting parks, zoos, museums etc. to see live animals and plants which cannot be brought into the classroom
- knowledge of living things to be broadened through books, charts, films
- observing and discussing how animals depend upon one another

About materials
The ideal curriculum would provide opportunity for:
- observing and exploring some materials at first hand displayed in the classroom
- discussing the meaning of words used to describe various properties (e.g. *floating or sinking, transparency, dissolving in water, hardness, flexibility, liquid, solid*, etc.)
- comparing materials in terms of some of their main properties, using 'fair' tests
- relating properties of materials to their use in everyday life
- relating some changes which occur in materials to the possible cause of these changes

Applying the criteria

The process of applying the criteria to the information about what is done is less complex than it may seem if carried out systematically. In effect what we have are two lists, one of items relating to what was done and the other of criteria relating to the major goals:

What was done	Criteria	
a	A	1
b		2
c		3
d		4
e		5
f	B	1
etc		2
		3
		4
		5
	C	etc

One method of proceeding would be to take **a** and look to see to which of the criteria it could relate, then do the same for **b**, **c**, etc. in turn. A preferred method is to work the other way round and to scan across the items under *What was done* to see if anything relates to A. If the answer is *yes* it should be easy to see to which of A1, A2, A3 etc. it relates. Clearly some criteria relate more to methods (e.g. perseverance) than to content and organization, whilst others relate more to content – as in the case of those relating to living things. When the information is arranged under these headings the scanning is made that much more easy.

The preference for this approach is that it soon becomes clear as to whether all the major types of goal are being catered for, thus it is a diagnostic approach which will give information for taking action. It also avoids the suggestion that what is done *only* has value in relation to one of the criteria, which is not necessarily the case. Certain items (**a**, **e**, for instance), may not meet any of the criteria but may still be worthwhile for other good educational reasons. The criteria do not in any sense present a basis for making absolute judgements as to what is worthwhile; they are only as useful and comprehensive as the thought which goes into producing them.

Evaluation in this context is not for the purpose of making summary judgements about whether the opportunities for learning are 'good' or 'not so good'. If this were the case there would be a question of deciding, for example, what proportion of the criteria should be met if the activities are to be considered satisfactory. If there were something being done relating to 50 per cent of the criteria for one goal, would this be enough? But it is not really productive to try to answer this question. The important use of the result of scanning the information in relation to the criteria is to bring attention to possible inadequacies in the opportunities provided. When it is known that little is done in relation to some criteria but a great deal that relates to others, the judgement as to whether this is a matter for concern and for action is one which only the teachers involved, knowing the pupils involved, can make. This judgement is probably best decided in discussion among teachers so that a common understanding can be reached as to what, in practice, constitute adequate opportunities *for their pupils* for certain kinds of learning. From this discussion it may become clear that, for instance, there should be a change of emphasis in teaching methods, more of one kind of activity and less of another should be planned for and, hopefully, that the evaluation should become a regular part of the teachers' responsibility.

The value of evaluation

The value of carrying out an evaluation lies partly in the foundation of evidence and analysis which it provides for making decisions about the curriculum, decisions which are otherwise made on a hunch, the latest new persuasion, by default, or are not made at all. But the process of carrying out the evaluation has value in addition to informing the decisions reached. This is especially so when groups of teachers in the same school attempt to define and apply explicit criteria to what they are doing. There may be differences between individual teachers, since criteria for evaluation are a reflection of views held about the purposes and processes of education. But it is better for these differences to be revealed and identified for what they are than for teachers to use conflicting criteria without recognizing that this is so. Often it is in the best interests of the pupils for the criteria used to be agreed by discussion – and compromise – among the whole teaching staff of a school than to be the idiosyncratic judgements of individual teachers.

References

APU, *Assessing Scientific Development at Age 11*, (DES, 1981).

Blyth, A. et al., *Place, Time and Society 8–13: an Introduction*, ESL Bristol for the Schools Council (Collins, 1975).

Brimer, A. et al., *Bristol Achievement Tests* (Nelson, 1969).

Deale, R.N., *Assessment and Testing in the Secondary School*, Schools Council Examinations Bulletin 32 (Evans/Methuen Educational, 1975).

DES, *Primary Education in England* (H.M.S.O., 1978).

DES, *Science in Schools. Age 11: Report No 1*, A report on the 1980 Survey of Science in England, Wales and Northern Ireland (1981).

DES, *Science in Schools. Age 11: Report No 2*, A report on the 1981 Survey of Science in England, Wales and Northern Ireland (1982).

DES, *Science in Schools. Age 13: Report No 1*, A report on the 1980 Survey of Science in England and Wales (1982).

DES, *Science in Schools. Age 15: Report No 1*, A report on the 1980 Survey of Science in England, Wales and Northern Ireland (1982).

Duckworth, E.R., *Evaluation of the African Primary Science Programme* (E.D.C., 1970).

Elliott, J. and Harlen W., *Portrait of a Project*, Final report of the Schools Council Progress in Learning Science Project. Available from the Schools Council Information Centre, 160 Great Portland Street, London (1980).

Ford Teaching Project, *Self-Monitoring Questioning Strategies*. Available from the Cambridge Institute of Education, Shaftesbury Road, Cambridge (1975).

Gall-Choppin, R., *Activities for Assessing Classification Skills* (NFER, 1979).

Harlen, W., 'Does Content Matter in Primary Science?', *School Science Review* vol. 59, no. 209 (1978).

Harlen, W., 'Selecting Content in Primary Science'. *Education 3 to 13* vol. 8, no. 2 (1980).

Progress in Learning Science Project, *Match and Mismatch. Raising Questions*, (Oliver and Boyd, 1977).

Kellington, S., *Assessment Questions for Integrated Science*, Book 1 (Heinemann Educational, 1979).

Kellington, S., *Assessment Questions for Integrated Science*, Teachers' Guide (Heinemann Educational, 1979).

Look! *Packs of Workcards and Teacher's Guide* (Addison-Wesley, 1981).

MacIntosh, H.G. and Hale, D.E., *Assessment and the Secondary School Teacher* (Routledge and Kegan Paul, 1976).

National Federation for Educational Research, *Record Keeping in the Primary School* (1980).

Prescott, G.A. et al., *Metropolitan Achievement Tests* (Harcourt Brace Jovanovich, Inc., 1978).

Rowntree, D., *Assessing Students. How shall we know them?* (Harper & Row, 1977).

Science 5/13, *Objectives for Children Learning Science*, contained in each of the 26 units for teachers (Macdonald Education, 1972–1975).

SCIS, *Life Cycles: Evaluation Supplement* (Univ. of California, 1973).

Straughan, R. and Wrigley, J. (Eds.), *Values and Evaluation in Education* (Harper & Row, 1980).

Tough, J., *Listening to Childen Talking* (Ward Lock Educational and Drake Educational Associates, 1977).

West Sussex Science 5–14 Scheme, *Science Horizons* (Globe Education, 1981).

Index

Activities, examples 1–3, 30, 31
Activities, recording 60, 61, 65
Assessment of Performance Unit (APU) 14, 41, 54–8
Attitudes, scientific 4–6, 8, 9, 24, 42, 51, 65, 67, 69, 71, 76

Blyth, A. *et al* 26
Brimer, A. *et al* 45, 48, 50, 51

Category marking 18, 28, 53, 54
Checklists 35, 58, 60, 61, 63, 65
Classroom organization 71–4
Concepts 5–8, 10, 24, 42, 49, 61, 63, 67, 71, 76, 77
Criteria, used in assessment 16, 17, 21, 28, 33, 36, 38, 39, 61, 63, 64
Criteria, used in evaluation 66, 67, 70, 71, 74–8
Criterion-referenced assessment 16, 18, 53

Deale, R.N. 14
DES, APU Science Reports 55–8
DES, HMI Primary Survey 22
Diagnosis 16, 22, 36–8
Duckworth, E.R. 38

Elliott, J. and Harlen, W. 38
Enquiry skills 4–6, 8, 9, 24, 32, 34, 42, 43, 46, 61, 65, 67, 71, 75, 76
Evaluation, meaning 66, 70, 71
Evaluation, curriculum 15
Evaluation, of records 61
Evaluation, and planning 75
Evaluation, time span 72

Facts, knowledge of 4–6, 42, 49, 61
Fair testing 2, 30, 46, 76, 77
Ford Teaching Project 37

Gall-Choppin, R. 34
Goals, primary science 5, 24, 27, 42, 75

Harlen, W. 5
HMI Survey of primary schools 22

Kellington, S. 42, 44, 46, 49, 52

'Look' 8

MacIntosh, H.G. and Hale, D.E. 14
'Match and Mismatch' 24, 25, 29, 62
Matching 22–4
Metropolitan Achievement Tests 47, 50
Multiple-choice questions 51, 52, 57

Norm-referenced assessment 17, 18, 53, 64
Objective marking 17, 52
Objectives, of science education 7, 8, 24

Practical tests 16, 17, 28, 33–5, 43, 46, 56, 58
Prescott, G.A. *et al* 47, 50
Process skills, *see* Enquiry skills
Profile recording 62
Progress in Learning Science Project 24, 29, 38, 61
Pupil-referenced assessment 17, 18, 28, 53

Questioning, by teacher 37, 73

Records, evaluating 66
Records, good 59, 64
Record Keeping in the Primary School (NFER) Project 59
Reliability 21, 27, 54
Rowntree, D. 14

Science Curriculum Improvement Study 33, 34
Science education 6, 36, 65
Science 5/13 2, 7, 8, 24, 69
'Science Horizons' (*see also* West Sussex Science 5–14) 42
Skills, enquiry, process, *see* Enquiry skills
Skills, of using information 32, 33, 42, 47–9, 65
Standardized tests 16, 19, 64, 65
Straughan, R. and Wrigley, J. 25

Teaching methods 71, 72
Teaching methods, evaluation 73
Tests, meaning 16
Test scores, recording 64–7
Tough, J. 37

Value judgements 25, 27, 36, 66, 71
Variables, in investigations 31, 32, 46, 76

West Sussex Science 5–14 (*see also* 'Science Horizons') 60
Work cards 8, 72
Written test questions 16, 17, 28, 43, 46, 51, 56

Science

Wynne Harlen

General Editor Professor Jack Wrigley

Acknowledgements

The ideas presented in this book developed over many years and through working closely with others. I am grateful to Jack Wrigley for the encouragement that I received when working at the School of Education, University of Reading, and the opportunity to bring together in writing this book some thoughts about assessment and about primary science. I would also like to thank those with whom I have worked during the past decade and who, by sharing ideas with me, have helped to produce this book.

© Wynne Harlen 1983

All rights reserved. No reproduction, copy or transmission of this publication may be made without written permission.

No paragraph of this publication may be reproduced, copied or transmitted save with written permission or in accordance with the provisions of the Copyright Act 1956 (as amended), or under the terms of any licence permitting limited copying issued by the Copyright Licensing Agency, 33-4 Alfred Place, London WC1E 7DP.

Any person who does any unauthorised act in relation to this publication may be liable to criminal prosecution and civil claims for damages.

First published 1983
Reprinted 1989

Published by
MACMILLAN EDUCATION LTD
Houndmills, Basingstoke, Hampshire RG21 2XS
and London
Companies and representatives
throughout the world

Printed in Hong Kong

ISBN 0-333-31916-8